THE BATTLE OF THE BEAMS

www.penguin.co.uk

THE BATTLE OF THE BEAMS

The Secret Science of Radar that Turned the Tide of the Second World War

Tom Whipple

bantam

TRANSWORLD PUBLISHERS

Penguin Random House, One Embassy Gardens,
8 Viaduct Gardens, London SW11 7BW
www.penguin.co.uk

Transworld is part of the Penguin Random House group of companies
whose addresses can be found at global.penguinrandomhouse.com

First published in Great Britain in 2023 by Bantam
an imprint of Transworld Publishers

A CIP catalogue record for this book
is available from the British Library.

ISBN 9781787634138

Typeset in 11.25/14.75pt Minion Pro by Jouve (UK), Milton Keynes.
Printed and bound in Great Britain by Clays Ltd, Elcograf S.p.A.

The authorized representative in the EEA is Penguin Random House Ireland,
Morrison Chambers, 32 Nassau Street, Dublin D02 YH68.

Penguin Random House is committed to a sustainable future
for our business, our readers and our planet. This book is made
from Forest Stewardship Council® certified paper.

For Mary and Raymond

Contents

List of Illustrations

Timeline

1939

1 September	Germany invades Poland. Reginald Jones takes up a post as Assistant Director of Intelligence (Science)
3 September	Britain and France declare war on Germany
1 November	Hans Ferdinand Mayer types up the document that becomes known as the Oslo Report
13 December	German pocket battleship *Graf Spee* is scuttled after the Battle of the River Plate
18 December	German radar spots a bomber raid on Wilhelmshaven. More than half of the twenty-two Wellington bombers are shot down

1940

21 March	Jones marries Vera Cain
8 April	Germany invades Norway
10 May	Churchill becomes Prime Minister
4 June	Dunkirk evacuation comes to an end
21 June	Jones meets with the cabinet, to convince Churchill of the existence of a German beam navigation system called Knickebein
5 July	An intercepted Luftwaffe message boasts of intercepting RAF flights using 'Freya'
10 July	Battle of Britain begins
1 August	Josef Kammhuber heads up a new organization tasked with coordinating radar defences of the western approach to Germany. It will become known as the Kammhuber Line

7 September	Blitz begins, with three hundred bombers raiding London
16 October	Ludwig Becker, a Luftwaffe pilot, shoots down a Wellington bomber at night after being directed to the target by radar operators on the ground
November	Charles Frank joins Jones
6 November	A Heinkel bomber equipped with the X-Gerät system crash lands on Chesil Beach
14 November	Raid on Coventry, directed by X-Gerät

1941

19 January	A Heinkel bomber is shot down. A notebook found on board gives key details of a new beam system, Y-Gerät
February	The RAF accepts the existence of German defensive radar
21 June	Thomas Sneum, a Danish pilot, makes a daring escape in a biplane, carrying radar intelligence
22 June	Operation Barbarossa: Germany invades Russia
8 August	Ludwig Becker again shoots down an RAF bomber using radar. This time, the radar is carried onboard his nightfighter
7 December	Japanese attack on Pearl Harbor, the USA enters the war

1942

11 February	The 'Channel Dash': two German battleships and a cruiser traverse the English Channel, in a crushing humiliation for the Royal Navy
15 February	Fall of Singapore to the Japanese

27 February	Operation Biting: the Bruneval raid
3 December	An RAF bomber detects Lichtenstein airborne radar being used by the Luftwaffe
December	RAF raid on a target in Florennes, using the Oboe navigation system

1943

30 January	H2S radar used for the first time, to provide bombers with a map of the ground below
2 February	Surrender of the German 6th Army at Stalingrad
24 July	First use of Window, in a raid on Hamburg
3 September	Italy surrenders

1944

6 June	D-Day: the largest amphibious invasion in history is protected by radio countermeasures and feints, deployed by sea and by air
25 August	Liberation of Paris
16 December	German counterattack, in the Battle of the Bulge

1945

| 7 May | Victory in Europe |

R. V. Jones

Prologue

'This was a secret war, whose battles were lost or
won unknown by the public; and only with difficulty
is it comprehended, even now'

WINSTON CHURCHILL, *THEIR FINEST HOUR*

21 June 1940

REGINALD JONES WAS LATE.

This was not an unfamiliar situation for the young scientist. Despite the pressures of the war, the 28-year-old Jones had pointedly maintained a more relaxed schedule. The first messages needing his attention at the Air Ministry near St James's Park rarely came in before 10 a.m., so why should he arrive earlier?

Unfortunately, however, this morning a message had for once arrived early at his office. On his desk, when he walked in at 10.10, there was a note from Daisy Mowat, his boss's secretary. It read: 'Squadron Leader Scott-Farnie has telephoned and says will you go to the Cabinet Room in 10 Downing Street.'

Jones and Mowat did not have a formal work relationship. Once, for instance, she told an important caller that Mr Jones could not come to the phone because he had jumped out of a window. He had to wrest the call from her, and – rather than deny he had climbed out of a window, which would have been an unsportsmanlike way to ruin a good joke – explain that that was just what he did for exercise. So he presumed that this message from Mowat too was a practical joke. As yet unconcerned, he went to check with her. She assured him it wasn't.

Jones rushed out and hailed a cab for the short journey to Downing Street. Then, twenty-five minutes after the meeting had been due to start, he pushed open the double doors and hurried into the Cabinet Room.

The meeting proceeded without interruption, and for Jones the sight of the dozen or so attendees must have been an intimidating one. Around the table sat Lord Beaverbrook, minister of aircraft production; Sir Archibald Sinclair, the Air Minister; Sir Cyril Newall, Chief of Air Staff; Sir Hugh Dowding, who led Fighter Command; Sir Charles Portal, who led Bomber Command; and Philip Joubert, in charge of RAF signals.

Then there were the scientific officers: the top brains in the land, who directed British research in this, the most technological war in history. There was Dick Tizard, who advised Air Staff; Robert Watson-Watt, who advised on telecommunications; and Frederick Lindemann – who was both scientific advisor and sage to Winston Churchill and, more pertinently, Jones's old professor. He was the reason Jones was here at all. Sitting in the middle of them all was Winston Churchill himself.

When Jones sat down, as quietly as he could, the conversation continued. The atmosphere was frosty, it was 'perhaps even that of a confrontation', Jones later recalled. There had indeed been a disagreement. Some of those present were unwilling to believe the claim, based on accounts from intelligence sources and half-heard conversations among downed German pilots, that the Luftwaffe had been using beams to direct their planes into Britain.

The startling theory was that the Luftwaffe could project two narrow radio beams over the country, forming a cross shape in the air. The German planes flew down the route of one of the radio beams, and when they met the other – at the point where the beams crossed – the bombers knew they had reached their target. It was an electromagnetic version of 'X marks the spot', and to many in this room it did not seem credible.

Some thought that such an idea was not merely fanciful, but impossible: a beam could not be that accurate. It could not, they pointed out, travel that far or curve around the horizon. A radiographic channel, carrying an armada of bombers to its target? It was barely worth considering. And why would any decent air force bother, in any case? British crews got by fine with stellar navigation.

Those gathered in that room were an impressive group, but as they argued Jones was not impressed. From the way they were talking, it was clear to him that many understood neither the science of the beams nor its importance.

At last Churchill turned to address a technical question to Jones. 'Would it help, sir,' the young man replied, 'if I told you the story right from the start?'

Writing after the war, Churchill described that moment as like sitting in the parlour while the great detective finally reveals the killer. 'For twenty minutes or more,' Churchill wrote, 'he spoke in quiet tones, unrolling his chain of circumstantial evidence, the like of which for its convincing fascination was never surpassed by tales of Sherlock Holmes or Monsieur Lecoq.'

The doubters were not necessarily won over, but they were no longer relevant. Jones, at the age of not even thirty, had spoken above their heads, to make an ally of the only man who mattered.

'Being master, and not having to argue too much,' Churchill wrote of that meeting, 'once I was convinced about the principles of this queer and deadly game I gave all the necessary orders that very day in June for the existence of the beam to be assumed, and for all countermeasures to receive absolute priority. The slightest reluctance or deviation in carrying out this policy was to be reported to me.'

The Battle of the Beams had begun.

At the beginning of the Second World War, much of the RAF establishment held two firm beliefs. The first was that through British pluck and ingenuity the country had, on its own, devised a superweapon: RDF, or radio detection finding, now known as 'radar'.

All along the south coast was a chain of powerful radio transmitters. These projected an invisible arc of radio waves out towards Europe. As bombers flew into range, the radio waves would ricochet and bounce off them, returning as faint echoes to the transmitters. Through these echoes, a central command centre could see the location of the Nazi planes and direct Britain's fighters to intercept them. Then, concentrating in a massed force on the invaders, they could – at least during the day – bring them down. The Germans, ignorant of this superweapon, would never know how they did it.

The second belief held by the British establishment was that German bombers, which relied on sight, could only be accurate by day – when they were vulnerable to attack by Hurricanes and Spitfires. Their British counterparts, on the other hand, would find their targets by night as well. Britain was a maritime nation, after all. Its people were trained in the arts of celestial navigation. At 20,000 feet, there is rarely cloud cover to prevent the use of astrolabe and sextant. The RAF bombers could become a nocturnal force, flying far above the sleeping citizens of Europe, evading anti-aircraft fire and enemy fighters under cover of darkness. The stars would guide Britain to victory.

There were hints, though, that maybe these beliefs needed to be questioned. The costliest hint came as soon as the British bombing campaign began – and a vast number of RAF planes started tumbling, fluttering and smouldering down on to the fields of France. In the years to come, desperate battles were fought over occupied Europe. Four miles above the ground there were heroic deeds and terrible sacrifices that will go forever unrecorded. But these individual tales, however brave, in aggregate became simply statistics, to be analysed back at the Air Ministry. And when they were analysed, what they showed was very unsettling indeed.

For each Bomber Command sortie, out of 100 planes that left their runways, 95 returned. Fly two missions, and your odds of making it to breakfast in the mess were roughly 90 per cent. Fly three, and they were similar to a round of Russian roulette. Just under one in six failed to make it any further.

Jack Furner, a navigator who went on to become Air Vice-Marshal, later recalled his period flying with 214 Squadron. On one six-month tour, in 1943, his plane was one of just two that survived. A 95 per cent return rate always had a different meaning to him. 'The figure says it clinically: to put it in a more meaningful way – on a typical operation involving 750 bombers, it meant the loss of 260 aircrew.'

Those numbers, each representing families whose sons, fathers and husbands would never return, told a technological as well as human story. To those who knew how to analyse them they showed the Nazis had radar too, and very good radar at that. The first of the RAF's treasured assumptions was wrong.

The evidence against the second assumption, that celestial navigation worked, was harder for the RAF to come by – although the Germans knew it almost immediately. This evidence came in the unarguable form of the bombs those RAF planes dropped, at such bloody cost. Far more often than not, rather than hitting their target these bombs exploded uselessly across the fields of Germany. A top-secret report in 1941 used aerial reconnaissance to estimate that of the planes that reported attacking their target, only one in three actually got within five miles of it. Aircrew were risking their lives to put craters in fields.

When Jones met Churchill in the Cabinet Room in June 1940, all this was to come. The Blitz had not begun, the British Expeditionary Force had only just been evacuated from Dunkirk. Misconceptions about British advantages still remained uncorrected. But among the small circle of men in that room the first chink had appeared. Because whether or not British pilots really could make it to Berlin at night by following the stars, it was clear that German pilots could do the reverse journey from Berlin to London without needing to. They just had to follow the beams.

It was 'a painful shock', wrote Churchill. 'The Germans were preparing a device by means of which they would be able to bomb by day or night whatever the weather . . . like an invisible searchlight [it would] guide the bombers with considerable precision to their target.'

With France about to fall, the first serious raids from the Luftwaffe were expected any day – to soften up Britain for invasion. As Jones sat down in that room, there was an obvious conclusion to be drawn. They needed to find the beams, and find them fast.

We know about Bletchley Park and the Manhattan Project. They have entered mythology. Just as conventional militaries understood that victory came from concentrating their army's firepower on a single target, these were places where a nation concentrated its brain power. Brought together into a fizzing critical mass of intellectual energy, these boffins – as the popular newspapers of the time called such men and (just occasionally) women – solved problems that seemed insoluble, and in doing so shortened the war.

There was another scientific war, no less important than code-breaking or atomic weaponry: the battle to outwit the enemy not merely in the air but on the airwaves. The full scale of this techno-logical war has often been overlooked, or if it has been recounted we have concentrated on just one aspect of it – the development of defensive radar. But the wider electromagnetic war was crucial first to survival and then to victory. The tools developed by its combat-ants allowed Britain – in its darkest hour – to see.

This war was fought with the tools to hand. The effort to beat the beams required jury-rigged countermeasures, designed on the fly and, on more than one occasion, commandeered in desperation. Somehow, though, this scientific war has never quite gained the same status.

Bletchley and Los Alamos were nurseries for the world's top scien-tists. Bletchley was where the finest Cambridge mathematicians applied their once arcane knowledge to codebreaking. The Manhat-tan Project was where the Allies' greatest physicists came together to solve the greatest practical problem in science. Their names would become famous, and sometimes infamous: Oppenheimer, Fermi, Bohr, Feynman, Fuchs.

Radio research was a similar process. After the war, released from their military duties, many of radio's leading figures would go on to populate the top universities. Some would stay in radar, developing the systems that could track Soviet ballistic missiles. Still others would move into parallel fields – building Britain's first nuclear power station or, in the case of Sir Bernard Lovell at Jodrell Bank, and Martin Ryle, the Nobel Laureate, adapting radar work to radio astronomy.

In the First World War the British army swept in on cavalry but soon saw their dreams of a decisive charge scythed down by the pitiless machine-guns of the Western Front. Likewise the notion of navigational exceptionalism did not survive contact with the enemy. Neither, though, in part thanks to Jones and his intervention at Downing Street, did the Nazi belief in their own mastery of radio waves.

Together these scientists, and the German adversaries they sought to outwit, created a new kind of war. At the start of the war the sky was a lonely place. A British bomber aimed by sight, and navigated using techniques that would have been recognized by participants at the Battle of Trafalgar.

By the end, there were navigation systems that had changed those lamentable bombing odds. They could direct a lone bomber to a single factory in the Ruhr – and leave it in rubble. There were radars that could spot the outline of Berlin from 20,000 feet through thick cloud. Meanwhile there were German anti-aircraft guns that could spot the bomber through thick cloud and shoot back. And there were also nightfighters that could spot all of these radio waves bouncing around and lock on to the bomber that sent them.

This is not the book to tell the history of all of these advances, or of all the people involved. Like Turing and Flowers at Bletchley or Oppenheimer and Bohr at Los Alamos, though, in the tale of this electronic one-upmanship there are names that stand out. One of these is Reginald or, to use the initials he preferred, R. V. Jones.

Jones is far from unknown. In the 1960s and 70s he emerged from the shadows of intelligence and briefly gained minor celebrity status – a status he rather enjoyed – with a TV show and an autobiography. Although that book retains cult status in the scientific community, outside it Jones has been almost forgotten. If the length of a Wikipedia page is a measure of a man's worth, his at the time of writing is 1,200 words. Turing has 10,000. Smoky, a Yorkshire Terrier adopted as a mascot by the US 5th Air Force, has 1,600.

This is an injustice. Jones was one of those rare scientists who had the ability to sit in a room and, with pencil and paper, see into the mind of his enemy. He worked out, often with just fragmentary information, what the Nazis were planning – and what the devices they built to achieve it would look like. Then he and others were able to devise ways to defeat them.

Unlike the caricature of the boffin, as exemplified by Turing, Jones was not someone who would avoid human contact. This young man, the son of a postman, was a practical joker who could look you in the eye, who would go to the pub with his most junior colleagues – but was unperturbed about contradicting the most powerful men in the land. He was a showman who would tell a good anecdote – even if after the war his contemporaries occasionally complained the stories were a little too good.

He was the most unlikely of leaders, but war is a great leveller. In a crisis, talent can rise. When the war began he was assigned to the Air Ministry as a scientific advisor. Because he needed a title, and it was far from clear what this job even was, he was made Assistant Director of Intelligence (Science). What he directed, and indeed who he assisted, was not clear. At the start, he was the only person in his team. Because he was very good at his job – a job he largely created – by his thirtieth birthday he was part of the small circle that knew the great secrets of the state.

Jones received Enigma decrypts direct from Bletchley; he had the ear of Churchill. At his request, pilots were sent on perilous photographic reconnaissance missions to German radar stations, and commandos on daring raids into enemy territory. He gained this power – or perhaps respect is a better word – not through his rank, but through his results, through an uncanny ability to predict what the Germans were about to do.

He was not working alone. There were others, many others, involved in radar and radio intelligence. There was the network of spies and resistance fighters in Belgium and France. There were the decoded messages and ingenious work of POW interrogations. There was the new science of photographic interpretation. And there was a team of some of Britain's most able scientists.

Then there were those who turned insight into activity. To be worth anything at all, what he and his colleagues discovered needed to be acted on: by the engineers and physicists who built the counter-measures, and by 80 Air Wing, the RAF unit who applied them. But, still, it was of Jones that Churchill would later say his 'magnificent prescience and comprehension . . . did far more to save us from disaster than many who are glittering with trinkets'.

During the four years after the retreat from Dunkirk the only Western Front that existed lay in the air. It was here that the Axis and Allied powers parried and probed. For every advance, there was a counter-advance. One side developed onboard radar on bombers to see attacking planes, the other put receivers on attacking planes to see the radar signals from the bombers they were stalking. Ground-based radar could spot individual bombers from distances well over 100 kilometres, but at the same distance individual bombers could also drop reflectors that deceived those on the ground into thinking they were a vast armada.

This was a game of shadows and feints in the skies above France. On any one night, flying over occupied Europe, there were new radar systems, old radio systems, and systems that existed solely to be jammed – decoys there so that the real advance they were hiding was ignored a little longer. Major technological improvements, that in peacetime would give an air force superiority for years, were intro-duced knowing that, at best, they had a useful lifespan of three months.

It would be decades before most of the public came to realize that during the Second World War air superiority – that term that became one of the defining strategic goals of the war – had also meant air-wave superiority.

And somehow at the centre of it was one man, a man half the age of most of those who he spoke over that day in Downing Street.

To understand how the young Jones had ended up here addressing the cabinet, to appreciate the chain of evidence that led him, like Lecoq, to his inevitable, inescapable conclusion, you need to go back. You need to go to a hotel room in Oslo in late 1939, where a travelling German businessman sat down to write a report.

PART 1

Defence

CHAPTER 1

The Killing of a Sheep at a Hundred Yards

'A beast like Hitler should not win the war'

HANS FERDINAND MAYER

Eight months before Jones's meeting with Churchill

ONE EVENING IN EARLY November 1939, a German businessman made a request of the head porter at the Hotel Bristol in Oslo. Would it be possible, he asked, to borrow a typewriter?

Hans Ferdinand Mayer was a middle-aged man with neat swept-back hair and the sort of boringly technical job that rarely invited further questions. As he took the machine up to his room on that cold Norwegian night, it seems unlikely anyone would have looked up from their drinks to watch a man in a generic suit heading upstairs. The chances are, in fact, that no one who saw him that day ever learnt that they had been witness to one of the greatest intelligence leaks in history.

Mayer was a brilliant scientist, who had become part of the German corporate machinery. He had studied under a Nobel laureate, published papers and patents, then risen through the ranks to head up the Siemens research laboratory in Berlin. Here, during the 1930s, his job had changed, subtly at first. Increasingly his role was not civil but military and, especially, he wrote, he found himself working on 'new and secret weapons which required communications and electronics techniques'. He understood all too well the significance. 'This meant – sooner or later – war!'

This worried him greatly. Mayer did not like the Nazis. He loathed them, in fact. He could not understand why his colleagues joined the Nazi party. '[Hitler] had proclaimed the Germans to be the "Master Race", superior to all other nations, which should work for them as slaves,' he said. 'He had also deprived the Jews of their citizenship, confining them to the status of "subjects".' It was not merely that this was, to him, abhorrent. Perhaps he could forgive his Nazi colleagues if they were merely evil. To his rational scientific mind, it was something else too, something far less understandable. It was, he said, 'such nonsense'. He wanted nothing to do with it. Later in the war, when he was awarded a medal for his work at Siemens, he took it home and pinned it to the pyjamas of his four-year-old son.

Even by 1939 he had already taken considerable risks in working against Hitler. A year earlier, he had learnt of a Jewish woman, married to an Aryan, who had been expelled. Because their daughter was half-Aryan, she could not leave with her mother. Because she was half-Jewish, her father did not want her. Mayer helped arrange for a passport for the girl under the name of an English friend called Cobden Turner, who effectively pretended to be her father.

Turner, who worked in the General Electric Company, had in turn pushed for Mayer to do more to oppose the Nazis. He had tried to get Mayer to pass secrets over to Britain. Mayer refused. He would only do that, he said, if the two countries went to war. Saying goodbye at their last meeting, in mid-1939, he had pressed a small and very powerful magnet into Turner's palm, part of an experimental device known as a proximity fuse.

Now, war had indeed broken out and the only way Mayer could save the soul of Germany was through helping the British win. But with the outbreak of hostilities his ability to contact Turner was gone. There was, however, another way. As the Siemens laboratory's director, Mayer was tasked with going abroad to represent the company in technical negotiations. This meant he was able to arrange his own travel – and he arranged it with care. On this particular occasion he was ostensibly in Norway because he needed to meet with a representative from the Norwegian Post shipping company. He was also there because, with four months to go until the German invasion, it was still a neutral country containing foreign embassies.

Closing the door of his hotel room and carefully putting on a pair of gloves to avoid his fingerprints being traced, Mayer began to type a report.

Being a senior scientist at one of the Reich's leading engineering firms meant that he was privy to information, or sometimes just gossip, from across German military research. It gave him a privileged position, in which he had an overview of the country's most secret projects. 'We worked together with the Army, Navy and Air Force, and there was hardly any secret weapon, conceived at that time, which was not known to me,' he later said. 'Some of these deadly weapons were already field-tested, and large-scale manufacturing had started.'

The report, written on two consecutive nights, reflected all that knowledge – and also its fragmentary nature. He started with a description of the bomber programme. Before the start of April, he said, 25,000 of the feared Junkers 88 aircraft were to be produced. He moved on to relaying Germany's plans for remote-control gliders and missiles. These would be particularly effective against ships, he suggested.

He had pieces of information, but not always the full jigsaw. Some of the sections of the report were way off, and obviously so to those who specialized in the subjects. Some were little more than a few sketchy sentences – one was about the name of a newly launched aircraft carrier, and the name he wrote down was wrong. Others were remarkably detailed. He revealed, for instance, that the Germans had an extraordinary kind of radio device. It was a prototype early warning system, he explained, that would bounce shortwave pulses off aircraft. Then – through dark or through cloud – it could display the reflections on a cathode ray tube. It had already spotted RAF planes from 120 kilometres away.

Another paragraph spoke of navigation, describing in detail a radio system that could tell a bomber how far it had travelled towards its target – providing precise information that could be used for pinpoint bombing. It suggested that German expertise in controlling radio waves was far further advanced than Britain had thought. It also suggested that, with a few extra tweaks – a narrow beam, say – they could guide bombers directly to their targets.

To one of the documents he pinned an actual proximity fuse, to be united with the magnet previously pressed into Turner's palm. Proximity fuses were small and unglamorous devices attached to anti-aircraft shells that would prove to be among the most important advances of the war. At the time, anti-aircraft fire was astonishingly inaccurate. Gunners not only had to aim quite literally miles ahead of their target to have a hope of hitting it, they also had to determine the correct altitude for shells to explode. Get either wrong, and the gun was useless. Proximity fuses were designed to make their job considerably easier. The idea was that they removed the guesswork – detonating in the presence of nearby metal objects.

At the start of the war, there were many different experimental approaches taken, by both sides, to achieve this. Some used miniature radar, with the shell bouncing a signal off nearby planes. Others contained light sensors that detected the shadow of bombers. The one gifted by Mayer was a German prototype, which was triggered by the tiny fluctuations in electric charge caused by the passing of large metal objects.

The two packages arrived separately at the British Embassy in Oslo on 2 and 3 November, landing on the desk of Captain Hector Boyes, the naval attaché. This was not in itself of note. The problem for Britain's embassy in Norway was too much espionage rather than too little. 'At that point,' confessed Boyes, 'one was inundated with various anonymous correspondence which it was necessary to sift.' This one seemed, he said, to have 'matters of interest'. He forwarded it on, and thought little more of it.

Mayer, doubtless, thought of little else. He was now a traitor, and had no idea if it had even been worth it. In the gloom of the northern European winter, he waited. At the bottom of the second report, he had placed a request. If it had been received, would the BBC be so kind as to change the greeting at the start of its World Service broadcast to 'Hello, hello, this is London calling'?

Later in the war, his habit of listening to the BBC would result in his arrest and imprisonment. Now, though, on the evening of 20 November 1939, at the start of the eight o'clock news, it brought instead validation. From across the North Sea came a crackly voice, the voice that in the grim years to come would remind him another Germany was possible. 'Hello,' it said, 'hello, this is London calling.'

Radio waves are just light you cannot see. Less than a century before Hitler invaded Poland, a scientist called James Clerk Maxwell realized that what we thought of as the spectrum was so much wider than the red through to violet we see in a rainbow. This, which we now call visible light, accounts for just the tiniest sliver of the smallest segment of what Maxwell called electromagnetic radiation. In the years since his death, we have learnt just how useful the rest of it can be.

Light is a wave, a form of energy that wiggles through space. When the wavelengths – the distance between a crest and trough – get shorter than 400 nanometres, you stop being able to see them and end up with X-rays. In 1896 Wilhelm Röntgen, the discoverer of X-rays, realized they could pass through some solid objects – even his wife's hand. When she saw the result she was shocked to see her bone structure clearly visible amid the fainter outline of flesh. She said, 'I have seen my death.'

When the wavelengths get longer, different properties emerge. After red comes infrared, which is visible to some insects (and special forces soldiers with the right equipment) and gets emitted by hot objects.

After infrared come microwaves, with wavelengths in distances that we can finally envisage on a human scale. Definitions differ, but typically the wavelengths of microwaves are anything from a millimetre to 30 centimetres. These too have proved useful. One day during the Second World War, an American called Percy Spencer was working on microwave emitters and he noticed the candy bar in his pocket had melted. He was intrigued. What if he blasted other foodstuffs? Popcorn popped. So too, rather satisfyingly, did an egg. The microwave oven was born.

As the wavelengths get even bigger, eggs stop exploding. But these larger, less frantic waves, known as radio waves, have an even more astonishing property. They can pass almost unimpeded through the atmosphere.

Between the wars, governments and the public came to the view that somewhere within this new electromagnetic spectrum lay a superweapon, waiting to be found. And they were pretty sure what it would be: a death ray. In the August 1924 edition of *Popular Radio*, there was an article titled 'The New Death-Dealing "Diabolic Rays"'. It is a measure of how passé such an idea was at the time that it did not even make the front of the magazine. The article meriting the cover slot concerned not diabolical death-dealing, but the rather more practical: 'How to build a two-tube reflex receiver.'

The lower billing of the death-dealing ray came despite the best efforts of the inventor behind it, a man called Harry Grindell Matthews. 'Has the dream of inventors come true at last?' the article began. 'Can destructive power now be sent through space by what has been called "the most terrible invention ever made by man"?' Then, in a journalistic phraseology that does for dubious technology what 'allegedly' does for dubious legal accusations, they added, 'Scientists disagree.' Grindell Matthews was in no doubt. He claimed his device, which used concentrated beams of electromagnetic waves, could ignite gunpowder, down planes, stop engines and even kill. And he said he had proved it with a mouse. 'As this beam of light crept up and finally struck full upon the little animal, he reacted exactly as he would have done to the shock of a light wire grounding through him. He was killed very quickly. He could have been killed instantly except for the danger of creating fire by applying the full force of the ray.'

Scientists did indeed disagree, and Grindell Matthews himself had a habit of never quite getting round to demonstrating his rodenticide technology when asked. But that does not mean that his ideas were not taken seriously. In all of the major powers, attempts were made to harness the electromagnetic spectrum to create rays of varying diabolical nature. Proposals included beams to boil the insides of enemy soldiers, beams to immobilize tanks, and beams to explode bombs while they are still held in the undercarriage of attacking bombers. Such was the profusion of claims that in Britain the Air Ministry offered a £1,000 prize for any inventor 'who could demonstrate the killing of a sheep at a range of 100 yards, the secret to remain with the owner'.

But eventually more sober-minded scientists than Grindell Matthews wrested control of the research agenda, and explained what was – and was not – possible. Asked by the Air Ministry to give his opinion on death rays in the mid-1930s, the British scientist Robert Watson-Watt – later present at Jones's meeting with Churchill – explained that he considered them unlikely to be a fruitful path of research. For one thing, radio is not a good way of transferring energy. Using a walkie-talkie does not boil your brain, and you hold your head right next to it. Visit the most powerful radio transmitters in existence, designed to broadcast television signals across the nation, and you do not see the grass around them scorched. It was easy enough to do the calculation to show that the amount of power required to raise a large animal's temperature by a degree – let alone boil it – was prohibitive. The sheep were safe. Equally fundamentally for those whose chief target is aeronautical rather than ovine, a metal aeroplane is shielded from electromagnetic radiation.

That did not mean the governments were incorrect, though. The electromagnetic spectrum does indeed contain a superweapon – and in his concluding reply, Watson-Watt hinted at what it might be. 'Attention is being turned to the still difficult, but less unpromising, problem of radio detection,' he wrote. 'Numerical considerations on the method of detection by reflected radio waves will be submitted when required.' In other words, maybe they should ask him about something more sensible than sheep-killing, and they might learn something useful. If it was a disadvantage, when it came to death rays, that a metal plane did not absorb radio waves, it was a huge advantage when it came to the scheme Watson-Watt was proposing. Imagine if instead of looking to fry enemy planes, you just wanted to spot them?

The suggestion came at just the right time. In 1932 Stanley Baldwin, three times Prime Minister, had warned in the House of Commons of the terrible toll a future war would take on civilians. 'I think it is well also for the man in the street to realize that there is no power on earth that can protect him from being bombed. Whatever people may tell him, the bomber will always get through,' he said. His speech scared and galvanized in equal measure. Today, it is often viewed as fatalistic. At the time, there was a slightly different understanding. A bit like with nuclear weapons in the Cold War, it suggested that the only way of protecting against a bomber force at night was to have your own bomber force. Deterrence was the correct, and only, response.

Not everyone accepted the inevitability of this doctrine, but none denied the problem. If Britain was to have any air defence system to speak of, then, above all else, it needed to be able to find the thing it was defending itself from. How do you do that on even a clear night, let alone a moonless and cloudy one? Anti-aircraft guns and fighter squadrons were no use unless they could find their targets. Frederick Lindemann stated that solving this problem was 'more important than a cure for cancer'.

At the time, the only system the country had to spot the metastasizing cancer of incoming enemy aircraft was the Observer Corps – a dad's army of civilian volunteers who took time out from their jobs as teachers and doctors and postmen to literally watch the skies. Theirs was not an organization that could reasonably be termed a fearsome military machine. One observer station was a telephone box in Wales, in which the telephone had been given an extension cable that meant its operator could stand outside and see the sky. In Scotland, a landowner grudgingly allowed the erection of an observation post on the understanding that its occupants maintained a 'decorous atmosphere'.

Today, radar seems the obvious and inevitable technological upgrade to decorous (or otherwise) volunteers in telephone boxes. That was not the case then. Back in the 1930s radar was merely one of the proposals on offer for preventing the bomber getting through, and not always the leading one either. If, well into the 1930s, you had asked Whitehall's boffins how we were to defeat the Luftwaffe's bomb-ers, they might instead have talked about 'Silhouette' – a scheme first thought up in the First World War that involved illuminating the clouds using searchlights. Looked down on from above, the clouds would be, in the words of one of Silhouette's proponents, a 'veritable carpet of light'. Then, circling at high altitude above, the fighters would wait, like prowling sharks. When they saw the shadow of a bomber passing over the clouds they would swoop.

Ask another boffin, though, and he might point to aerial mine-fields as the solution. Just as the ports could be protected by floating mines and the beaches with buried ones, what if we could do the same in the sky? When a raid was detected, rockets would be fired in its path, each pulling behind it a wire strung with small bombs, like malevolent fairy lights. Any plane flying into a wire would become snagged, getting tangled in a mess of explosions.

A third scientist might give a different answer still. Yes, we could use waves to detect incoming planes – but surely sound waves, not radio waves? Today, at Denge in Kent, you can see the results of this idea: vast polished concrete disks, 'acoustic mirrors' that were to be part of a string of listening stations all along the south coast. It was only when they were tested in a full-scale exercise that the decision was made not to build more. For planes travelling at 300mph, it was found that a system that relied on receiving information at the speed of sound – just over double that speed – did not offer enough warn-ing. By the time the alarm was sounded, there was just ten minutes to get planes to the target. In the exercise, first the Air Ministry was noted as having been destroyed, then the Houses of Parliament. What is more, if the acoustic mirrors failed to raise an alarm when needed, they also had a habit of doing the opposite when they weren't. The whole system could be jammed by a milkman clanking by in a horse and cart, then whistling while on his rounds.

Perhaps radio waves – rarely produced by milkmen – offered bet-ter hope than sound waves after all? A little after Watson-Watt's memo a Heyford bomber flew, as arranged, past the BBC short-wave radio station at Daventry. On the ground, Watson-Watt had impro-vised a system to see any reflections of the BBC signal. If the powerful radio signal pumping out of the Daventry transmitter was at any point returned, if the ghost of a reflection was detected, it would be seen as a beat on a cathode ray tube – a device that could turn electrical signals into visual ones, most notably in pre-flat screen televisions. As the ponderous aircraft, the last biplane bomber to serve in the RAF, flew over, there was a pulse that lasted several minutes. It was a single electronic flicker on a single screen – the faint energy returned from a massive radio transmitter. Judging direction, range, and height was all but impossible; but in that bouncing light, in that one experiment, lay the device that would – as much as any other – protect Britain in the war to come.

Radio direction finding, later known as radar, was born. Watson-Watt was ecstatic – they had a way to protect the country's approaches. 'Britain,' he said, 'has become an island once more.'

By 1939 a protective and invisible net was cast around the country from Southampton to Sutherland. Sites were chosen for their unobstructed views into Europe, for their good electrical access and at places that would not 'gravely interfere with grouse shooting'. They were switched on just in time to catch the approaching Luftwaffe. Then, in the skies above Britain, plucky fighters could be sent to where they were needed. As Churchill later said, never in the field of human conflict was so much owed by so many to so few. And those few – those young pilots in the skies above England – would have been useless without the invisible hand of radar that guided them.

That, at least, was the British story, the tale of how the nation developed a top-secret technology that would confound its enemies and save it in its darkest hour. But one country does not have a monopoly on dogged ingenuity and inventive pluck. At almost precisely the same time as Britain was congratulating itself on its prototype radio superweapon, Germany was putting the finishing touches to its own, rather more sophisticated radar. Whereas the first British versions simply told you an aeroplane was nearby, the German radar did so much more. It could spot a ship at 8 kilometres away and fix its bearing and distance with an accuracy of 50 metres – enough to direct artillery through radar signals alone. When scanning the air it was, they discovered by chance, just as impressive. During testing, a seaplane flew over, 28km away, and produced a clear reading.

It had its detractors, nevertheless. The head of technology at the Luftwaffe was very suspicious, if only on the grounds, he said, that once radar was incorporated into planes themselves, 'flying won't be fun any more'. Hermann Goering tended to sympathize too. 'Radio aids contain boxes with coils and I don't like boxes with coils,' was his unanswerable objection. But if German radar did not assume the near-doctrinal importance by 1939 that it had in the RAF, that was only because it did not represent the kind of war Germany was planning to fight. Hitler's forces were not going to be entrenched at the German border, waiting for planes to attack, as Britain's were. Theirs was not to be a static trench war, like the one a generation earlier. Germany's air force was instead going to be pushing forward, onwards, in a Blitzkrieg that would overwhelm the enemy, and its radar, without the need to consider their own defence. For that they needed to know where their own planes were – not the enemies'. They needed roads in the sky, taking the Luftwaffe to its targets.

As the document that Mayer took such risks to type over those evenings in Norway showed, German radar technology was easily comparable to that in Britain. More than that, though, the Oslo Report – as it became known – implied that as well as radar the Germans had used electromagnetic waves in a different way: to guide their own planes rather than reveal the enemy's. They would take the Luftwaffe, the tip of the Blitzkrieg spear, to their targets with absolute precision, whatever the weather or time of day.

The Oslo Report should have been a warning of the battle of the beams to come, of the folly of believing that this would be a war in which twentieth-century airmen navigated like eighteenth-century seamen. In writing it Mayer had given the British information that could save thousands of lives – and that also, if discovered, would end his. But to be useful, someone first had to take it seriously.

Hans Ferdinand Mayer was arrested in 1943. It was inevitable: he had become too much of an annoyance. He smuggled valuables abroad for Jews; he was overheard criticizing the Nazi regime. Not only did he mock the medal he had been given by putting it on his son's pyjamas, he then passed around photographs of it. He was, the Gestapo believed, grit in the oil of the smooth Nazi machine.

In the end, it was the BBC that did for him. A maid in the house next door overheard him repeating something that had been said in one of the broadcasts, and she went to the Gestapo. To survive, he needed the help of a genuine Nazi. His wife appealed to Philipp Lenard. Lenard was his former doctoral supervisor but, more importantly, he was an avowed antisemite who was 'Chief of Aryan Physics' for Hitler and considered Einstein's relativity a 'Jewish fraud'. Lenard wrote back to say, 'Your husband must change his mind. It will not harm him if he thinks as I do. Misfortune is the soap God is washing us with. We cry like little children, when they are being washed.' He did, however, promise to intervene with Himmler.

Mayer was lucky. He was sent to a concentration camp, rather than executed. Although the Gestapo had discovered his secret habit of listening to BBC radio, they did not know that through him their own – far more consequential – radio secrets had also been discovered. No one ever learnt of the night he listened to the BBC as it sent a special message just for him, a message that confirmed one of the greatest leaks in the history of military scientific intelligence. He was also lucky because he was still valuable. Perhaps because of Lenard's intervention, in Dachau he worked in a radio research laboratory, staffed by inmates. Ironically, some of its work would involve counter-intelligence against Britain – trying to undermine the very team his leak had hoped to help. The work kept him alive. He had privileges, food and relative safety. He learnt to, in his own analogy, 'emulate a mouse' – to be as unobtrusive and uninteresting as possible. In the final days, as the regime collapsed and order deteriorated, he found an opportunity to simply walk out of the camp. He kept walking until, amid the forest, he came across a cottage. There, he saw a mighty Nazi eagle carved from wood, and a mother, newly bereaved, hacking it with an axe in fury. Here he had an ally, and she gave him food and a place to sleep.

He had survived, and kept his secret. In fact, it would be decades before any Germans found out. After the war, amid the ambivalence felt towards those who betrayed their countrymen, he wanted to remain anonymous until after his death.

It was almost for nothing. His report should have changed everything, but the British were suspicious, and with good reason. On the one hand, the document they received made elementary errors and contained irrelevancies – such as a detailed and idiosyncratic section on attacking Polish fortifications. Yet on the other hand it offered intelligence that was too good to be true. How, MI6 wondered, could one man really know so much about such a broad range of weaponry? In the game of bluffs and double bluffs that have always blurred the shades of grey of espionage, the British intelligence assessment was that the document was a plant. If the disclosures were to be believed, the Nazis not only had their own radar and radio-guidance, they were also developing implausibly advanced superweapons including flying bombs powered by devastating rockets. How could one man have all this information? No, they decided, this supposed leak was designed to deceive, to have them fighting shadows. As one official report later ruefully put it, so the logic of the time went, it was 'planted by a wily enemy', who made sure that the few checkable facts were true – so that we would foolishly believe all the rest.

On this basis they all ignored it. Except, that is, for one young intelligence officer. Later, this officer said he came to treat it like a guide to the future. 'In the few dull moments of the War I used to look up the Oslo report to see what should be coming along next.' This man was Reginald Jones.

CHAPTER 2

Reginald

'In the field of radar they must have the world's greatest geniuses. They have the geniuses and we have the nincompoops'

HERMANN GOERING

HAROLD VICTOR JONES WAS one of the 'Old Contemptibles', the shattered rump of the British Expeditionary Force that was sent to hold the line in 1914, and was almost wiped out over the four tough years that followed, fighting desperately in the trenches of the Western Front. He was lucky to have made it through, and very nearly didn't. His left arm had been smashed, the result of a desperate battle at Neuve Chapelle – in which sixteen out of twenty-one officers present died.

Now, the war was over and the eight-year-old Reginald Jones and his mother were in London for a victory parade. As Harold Jones and his comrades marched past, the crowd, confused, fell silent. They were wearing suits. Who were this group of men marching in civvies? After the war ended the army had been operating on a 'first-in-first-out' policy. This meant that Harold and the scattered remnants of that first army had already been demobilized and were now out of uniform. Then, the silence was broken by a call from

Reginald's mother: 'Cheer the men in civvies, they were the men who went first!' The crowd erupted.

Jones did not come from poverty exactly – but the world his father created for him on returning from the Western Front was definitely not one of wealth. Much later in his career, when he began moving in more elevated circles, he confessed he resorted to reading a book of quotations – as a way of rapidly filling the hole he perceived in his education.

The family lived in Dulwich, south London. There, after the war, his father had put his two mismatched arms to use as a postman. Jones was educated at Alleyne's, then a solidly middle-class school. Occasionally, for fun, he would tinker with radio sets – an activity that was for his generation of schoolboy geeks what programming would be for those born in the 1970s. 'There has never been anything comparable in any other period of history to the impact of radio on the ordinary individual in the 1920s,' wrote Jones, of his passion. 'It was as near magic as anyone could conceive, in that with a few mainly homemade components simply connected together, one could con-jure speech and music out of the air.' One set he built in 1928 was so sensitive that he conjured Australian speech out of thin air. The Mel-bourne radio station that had originally broadcast it ten thousand miles away, never imagining that it would be able to curve round the world, sent him a prize in acknowledgement. It was not the last plau-dit Jones would receive for realizing that radio did not always go in clean straight lines, that it could bounce and bump off the upper atmosphere.

Throughout his youth, Jones could never quite escape a feeling that the twenties and thirties were not the end of the global turbu-lence, as many thought, but merely the passing eye of the storm. At school he remembered the headmaster Mr Henderson warning about the Germans, and how they could not be forgiven for the First World War because they had never expressed regret. 'Mark my words, as soon as they're strong enough they'll be at us again!'

Jones always claimed that he chose Oxford because of an incident at school. It had been the annual university boat race, followed closely throughout the country, and he needed to pick a side. 'However partisan the undergraduates of Oxford and Cambridge might have felt about the outcome, they were almost as conscientious objectors compared with the belligerent boys of the typical London school,' he said. One of those belligerent boys asked him which side he supported and Jones, thinking quickly, replied 'Oxford'. Wrong answer. 'He promptly punched me on the nose.' He would be Oxford ever after. Whether that is strictly true or not, the fact was that he did go to Oxford, and it was a fork in his life. Here he met arguably the two most important people in his future career. The first was Charles Frank, and the pair bonded over gassing Oxford.

It had been Charles Frank's idea, but Jones was a more than willing accomplice. The plan was simple. It was Guy Fawkes Night and, Jones later explained, in a tale that may just have gained a few embellishments for the benefit of his audience, it was traditional that 'there should be a "Rag" in the streets of Oxford'.

The most exalted prize on such a night was a policeman's helmet, ideally deftly removed from a policeman's head. To reach the state of chaos in which policemen were both plentiful enough and preoccupied enough for students to have a shot at such a prize generally required some kind of distraction. As it was 5 November, that distraction was, typically, pyrotechnic.

Jones and Frank had a different plan. Not for them mere explosives. Instead, they were going to 'out-do the raggers with their fireworks by introducing chemical warfare'. In his studies, Frank had been working with a chemical which 'turned out to be a virulent lachrymator' – a tear gas. It was perfect for the job in hand. Carefully secreting it in vials, they went out and, at the appointed time, smashed them on the ground. And nothing happened.

Vial after vial was smashed, yet the crowds continued to mill, unperturbed. It was, they later realized, too cold for the liquid to evaporate. Instead, people trod on it and carried it around the city on the soles of their shoes. There it slowly warmed and dispersed until, suddenly, it seemed like the whole city was crying. Students clattered along the cobbled streets, running for the safety of their colleges. Townspeople, their eyes streaming, were hauled into ambulances. The crews of those vehicles, some of whom almost certainly had memories of the First World War, ordered the crowds to disperse. 'Mustard gas!' they called, 'Clear out!'

Jones and Frank had the chaos they wanted but, amid a fearful mob and veterans reliving the horrors of the trenches, it did not seem quite like a helmet-pilfering atmosphere. In fact, Jones noted, such an atmosphere did not return the year after or the year after that. 'It seemed to break the whole tradition . . . things were very much quieter in succeeding years,' he recalled.

The second of the crucial Oxford contacts Jones made came about through very different circumstances.

When still an undergraduate, he sat an end-of-term examination, which came in two parts. The first part involved challenging problems, the second more conventional questions. Jones got so interested in the first set that he forgot to move on to the second until only fifteen minutes were left. The examiner, on receiving a somewhat lopsided paper, didn't care. He said he had never seen his questions answered so well (notwithstanding the fact he had also never had someone spend so much time on them), and already had him marked down as fellowship material. That examiner was Frederick Lindemann, then in his forties – a brilliant but arrogant physicist who had a tendency to make firm friends and even firmer enemies. Jones, luckily for him, fell into the first camp.

It was not just physics that the pair had in common. The early 1930s were, certainly for Jones, a strange time to be at Oxford. This was when the Union held the infamous debate in which it passed the motion, 'This House would not in any circumstances fight for King and Country.' It is difficult, today, to understand the effect it had – and the significance given to what was a glorified student debating society. Mussolini was said to have been heartened by it, and Hitler too. Churchill, at the time a continual irritation to the government, bloviated from the backbenches like a port-soaked walrus. He called the vote 'abject, squalid, shameless' and, for good measure, 'nauseating'.

Jones agreed. For him, it was naive and disgusting. He felt he was the only one of his contemporaries worried about the rise of Hitler. He was so ashamed that he wrote a letter home to his mother, apologizing for the behaviour of his university. 'I told her not to judge Oxford by the aspiring politicians in the Union, and although most of my colleagues were at that time pacifists, I thought that many of us would fight.' A war was coming, he was increasingly sure, and it would be a scientific one. Jones wanted to be there, fighting for King and Country – not part of a decadent intelligentsia of posturing conscientious objectors.

On this, he found he had an ally in Lindemann – and through his professor a connection also to Churchill himself. Lindemann and Churchill had been friends since the end of the First World War. They were an unlikely pair – Churchill the whisky-drinking, cigar-smoking meat eater, Lindemann the teetotal vegetarian. But their bond was such that when, in the House of Commons, an MP had questioned Lindemann's influence, Churchill had muttered, 'Love me, love my dog, and if you don't love my dog you damn well can't love me.' In 1932, Lindemann and Churchill had gone on a road trip through Europe, and what they saw there appalled them both. 'A terrible process is astir,' said Churchill afterwards. 'Germany is arming.' Lindemann agreed.

Here, at least, was someone in Oxford who shared Jones's concerns. He was also someone who was actively working to address them. The pervasive attitude that, in the words of Baldwin, 'the bomber will always get through', made Lindemann so cross that he wrote a letter to *The Times* on 8 August 1934: 'That there is at present no means of preventing hostile bombers from depositing their loads of explosives, incendiary materials, gases or bacteria upon their objectives I believe to be true; that no method can be devised to safeguard great centres of population from such a fate appears to me to be profoundly improbable.' Assuming otherwise was, he argued, 'defeatist' – and meant accepting that 'bombing aeroplanes in the hands of gangster governments might jeopardize the whole future of our western civilization'.

Well, Lindemann was not going to have it. On his return from Europe, with Jones's undergraduate studies at an end, he offered

Jones a position, which he took up – working in Oxford's Clarendon laboratory looking at the detection of infrared light.

Even in the late 1930s, it was not clear which bits of the electromagnetic spectrum would provide the next superweapon, nor what form it would take.

Electromagnetism is light. To make light of any kind you accelerate electrons, the subatomic, almost mass-less particles that orbit the heavy nucleus of an atom. If a hot poker is glowing red, it is because the electrons are bouncing around so fast that the wavelength of electromagnetic radiation they produce is the same as that of the light we call 'red'.

One way to move electrons is to heat up objects – causing their atoms to jiggle. Another is to create electricity, moving the electrons between atoms. In plugs, electricity runs on '50 Hz alternating current'. This means that fifty times every second the current in a house changes direction. Since current is just the flow of electrons, this means the plug is accelerating electrons – forward and back, forward and back – and so generating electromagnetic waves.

It is easy enough to work out the wavelength – the distance between two crests of a wave – of the light made by accelerating electrons in your plug. Each time the electrons are swished first one way then the other, a wave of photons of light head off – remember, radio waves are just invisible light. They depart on their journey from your plug socket, naturally enough, at the speed of light: 300,000 kilometres per second. And this happens fifty times a second. This means that after the first swish the light created by your plug has travelled a fiftieth of that – or 6,000km. That is your plug's wavelength, and that is also the reason why your house's circuitry does not interfere with your radio set: its wavelengths are many, many times too big.

One of the great challenges of work in the electromagnetic spectrum is to develop the devices that can jiggle electrons reliably at the right frequency to produce the wavelengths you need. Microwave ovens jiggle electrons 2.4 billion times a second, meaning that they produce a wavelength of about 12 centimetres. X-rays jiggle a billion times faster still, producing incomprehensibly small wavelengths.

For Jones's first project, though, this was not an issue. Lindemann tasked him not with making electromagnetic radiation, but receiving it. He asked Jones to make infrared detectors – devices that converted this light, with a wavelength just below that picked up by our eyes, into visible light that people could see. It was an intriguing project. What mysteries of nature, hitherto hidden, would become apparent in infrared?

Some mysteries did indeed become apparent. For Jones's PhD, he investigated the infrared absorption of coloured minerals. When he got really good at making detectors, he would point one at a row of metal tools, ask someone to pick up and replace a metal spanner while he was out of the room, then tell them which they had touched just by spotting the residual heat.

By 1935, however, it was fairly clear that if infrared was to have a role in the next few years, it would be neither in mineralogy nor party tricks. All science, as Hans Ferdinand Mayer had found in his own work in Germany, was now orienting towards war. Jones was sufficiently aware of its potential national security value that when a German turned up unexpectedly in his laboratory, apparently behaving like a comically B-movie Nazi agent, he realized he had to organize a diversion.

By the time Hans Wilhelm Thost arrived in Jones's laboratory, MI5 had been intercepting his mail for three years. It is a measure of Thost's level of skill as a Nazi secret agent that he did not notice this. Early on in his posting to Britain, he had even gone to enquire at the post office why it was that, for reasons he could not fathom, his foreign newspapers – which didn't need to be steamed open and inspected – always arrived on an earlier post than his foreign mail. The agent watching Thost was cross, accusing the post office of 'dropping a brick'. He instructed them to hold back the newspapers as well. As far as they could tell, once all his mail was delayed Thost never suspected a thing.

Thost was ostensibly in Britain as a newspaper correspondent. His role as a spy was not so clearly defined, but seemed to be directed less towards cloak and dagger than port and leather-backed chairs. 'Thost has instructions to gain access to London clubs and to create a pro-Nazi atmosphere,' concluded the security services in 1933, shortly after Adolf Hitler's election victory. 'He has also been told to get in touch with the better class younger generation and to pay special attention to university students.'

The closest he got to a genuine intelligence coup was when he recruited Christopher Draper, a First World War flying ace and minor film star. Draper had a habit of flying under bridges, which had earned him a nickname – the 'Mad Major' – but not much else. Draper was, as MI5's report stated, in dire financial straits – or, as they put it, 'on his beam ends'. Security officials strongly suspected he had considered the German offer, which included payment, ser-iously. But his patriotic instincts – or self-preservation – won through, and he reported the encounter to MI5 immediately. He spent much of the meeting trying to ascertain whether, if employed as a double agent by the British, he would be allowed to keep any money given to him by the Germans.

So it seems reasonable to conclude that when Thost met Jones he was still looking to impress his handlers back home. Jones, who was arguably not among the 'better class' of students, was not his reason for visiting the Clarendon laboratory. Thost had come to visit a Ger-man researcher – but he wasn't there. While they tried to find him, Thost waited in Jones's office. This concerned Jones, who didn't wholly trust his visitor around his infrared detectors. 'A newspaper correspondent might easily be a cover-operation for a spy,' he noted. So he decided to distract him.

'I told Thost that I had a certain amount of sympathy with Hitler, and could see why he had pushed out the Jews. Thost almost clicked his heels together, with an "Ach, so!"' Yet, Jones said, regretfully he feared the expulsion policy might backfire. 'They are very clever,' he said, 'and if they started to plot against Germany there could be trouble. For example,' Jones added, faux-casual, 'I know there is a great anti-Nazi organization run by the Jewish refugees in Britain.' Thost responded, Jones said, 'With a highly sibilant "Sso!"' and started to take notes. '"Oh yes," I went on, "I thought everyone knew about it. Why, the headquarters are here in Oxford!"' At this, Thost announced he could wait no longer for his friend, and had to rush back to London. A few weeks later, having presumably passed the details of this totally fictitious organization back to Berlin, he was finally expelled from Britain.

For Jones, the encounter with Thost was clearly an amusing diversion. But it was also a sign that his work and life were intersecting ever more closely with national security. There was no pretence any more that he was conducting pure research into the electromagnetic spectrum – using infrared detectors to uncover the mysteries of a hitherto invisible world. By the end of 1935 he had pivoted entirely to seeing if they could be used to spot planes by detecting the heat of their exhausts or engines. He was still at the Clarendon in Oxford, but now the government was paying his wage – he had become part of the Air Ministry. It was work that was ahead of its time. Perhaps so far ahead that it was essentially pointless. But amid the myriad of different ways to prevent Baldwin's unstoppable bombers from always getting through, it was no more implausible than aerial mines or floodlit clouds. Throughout the late 1930s, Jones worked on the problem, but by 1938 it was clear it was going nowhere. Now firmly on the government payroll, he was sent to Teddington, to work for the Admiralty's scientific service – where he was bored and aimless.

Leaving academia also had significant downsides in and of itself. As it became clearer that infrared was not going to be the technology of the future, Jones became depressed. Wasting years of your life on research was one thing; wasting years of your life on research you couldn't even publish – that remained top secret – was quite another. It was, he said, 'a rotten reward for three years of desperate work, from which I could not even recover the kudos of papers in scientific journals'. He would have seriously considered leaving, except that he needed the money. His father had lost the vision in one eye, and feared the second could go too. Jones realized there was a good chance he would have to support both his parents.

His personal gloom mirrored, for him, the national picture. As Europe reached crisis point, many were relieved by the apparent victory of Neville Chamberlain's diplomacy. Jones was not. When

Chamberlain flew to Munich and returned holding a 'pathetic scrap of paper' announcing 'Peace in our time', Jones felt isolated, among an intelligentsia almost universally supportive of the agreement. He saw it as a sham – a way of duping Britain into peace so that Germany could wage war, later, on its own terms. It merely 'postponed the reckoning'. He also felt useless. 'I was angry as a cat which had just been robbed of its mouse,' he said. He didn't know it, but as Chamberlain's plane completed its historic flight it was tracked by a new kind of device, a device that could ping out radio waves at the speed of light and then watch for the faint return of their reflection – a device that, unlike infrared, travelled almost unimpeded through the atmosphere.

By the time Chamberlain's scrap of paper was indeed shown to be pathetic, with Germany invading Czechoslovakia, Jones was still in the Admiralty, and kicking his heels. The country though, at last, was not. Amid the panic of the preparations for war, from one of the Whitehall committees that had been hastily convened there came the realization that there was a gap in Britain's scientific knowledge. Although the country had great scientists working on its own technology for its own defence, it had no scientists working on understanding German technology. As an experiment, a new role was mooted. Could a young scientist be seconded to the Air Ministry, to conduct 'scientific intelligence' – to coordinate and liaise, to provide a more technical eye, in a department more used to employing classicists and historians? It was, in many ways, a non-job. But Jones saw in it the potential to change the war, as 'one of the very outposts of our national defences'. In his view, 'A failure to detect the development of a new German weapon could easily lead to disaster.' Or, as he put it when accepting, 'A man in that position could lose the war – I'll take it!'

He also realized how badly such work was needed. He, like many, had had an idea of the secret service as some kind of panopticon, a super-competent all-seeing organization defending our democracy with all the resources of the state. His first trawl through its technical documents quickly disabused him of that notion. It was amateur, haphazard, and particularly poor when it came to science. 'My search through the files taught me how primitive was our intelligence service compared with what, from a schoolboy onwards, I had imagined it to be.' It needed a guiding hand.

He had got his commission, a sense of purpose, and his own front line. 'As with my father in 1914, I would hope to be among the first.'

CHAPTER 3

Reading the Tea Leaves

'We need not feel ashamed of flirting with the zodiac.
The zodiac is well worth flirting with'

D. H. LAWRENCE

Britain, 1940

THERE WAS TO BE a 'symphony of death', said Louis de Wohl, a
German-born writer employed by MI5. 'From all parts of the
world,' the naturalized Englishman told Britain's top intelligence
chiefs, 'our instruments must play the melodies.' In Cairo, where
Rommel still threatened to break through into the Middle East ter-
ritories, De Wohl declared that the local newspaper would proclaim
the imminence of Hitler's death. In the USA, still neutral, there
would also be reports that the German leader's days were numbered.
De Wohl was, in fact, prepared to employ his entire network of 'intellectual powers in foreign countries' to sing the same tune – each of
them on his orders would place predictions in print of Hitler's death.
Thanks to De Wohl's contacts, this requiem march for the Führer
would be heard across the empire and beyond. Provided, that is, you
paid attention to the horoscopes. De Wohl's 'intellectual powers', his
network of contacts, were astrologers.

De Wohl was, in the words of one intelligence officer, 'a tall, flabby elephant of a man'. He was an opportunist, a crank but – probably – not a charlatan. He appeared to be a true believer in the power of astrological projection. And, in the early frantic months of the war, when in villages across Britain the forearms of church bellringers were primed and ready to sound the chimes of invasion, he had the ear of government.

The idea of seconding an astrologer into the intelligence service was not as completely absurd as it might sound. The theory, at least as described by Admiral John Godfrey, head of naval intelligence and the man who brought De Wohl in, was that it was not necessary for astrology to have any validity for it to nevertheless be useful. 'Under certain circumstances,' said Godfrey, 'it is what people believe that matters, not what is.' His intelligence colleagues could think that astrology was nonsense; they might even be right. Hitler, they believed, did not. 'It should be observed that astrology is claimed to be an exact science and that given the hour, date and place of birth . . . all reputable astrologers will arrive at roughly the same conclusion,' Godfrey said. If true (and that's a big if), this meant that De Wohl could interpret the astrological charts of the upper echelons of the Axis powers, and through those charts the upper echelons of the Allied powers would hear the advice their opposite numbers were receiving.

So it was that one of the more unusual reports in British military history came to be circulated to, among others, Sir Dudley Pound, the First Sea Lord. It was titled, 'The Astrological Tendencies of Herr Hitler's Horoscope, September, 1940 – April, 1941'. It answered the biggest question of the day. When would invasion come? Hitler would not, the report assured its readers, move in September, as the 'directions' were unfavourable. They remained unfavourable in the start of October, as did the 'lunation'. So bad was Hitler's horoscope then, in fact, that De Wohl advised that Britain's depleted and demoralized expeditionary force launch a counter-invasion, because 'he knows that luck is against him'. Come December, though, Hitler's planets are at last aligned, and, he explained in the report, we should fear the launch of his invasion barges. More than that, at that point the Italian leader's planets are not aligned. Here, De Wohl voices a suspicion. He noted that Herr Hitler 'always asks Mussolini for a conference when the Italian leader is badly aspected'. His interpretation? He wants his ally to fail almost as much as he wants himself to win – so that only one power dominates Europe. In December, De Wohl predicts, the German army will prepare to deliver a crushing blow to the British Isles at the same time as luring the Italians into a futile attack on Britain in North Africa, 'so as to humiliate him through defeat'. Britain will save face by defeating Italy, then sue for peace with a victorious Germany.

The reception given to this report was not, it has to be said, unanimously positive. On the front page, in neat handwriting, Pound writes, 'Interesting, but I should like to work on something more solid than horoscopes.' It was not a complete dismissal, though – and encouraging enough that De Wohl's employment continued, allowing him to develop his far more ambitious plan to take astrology on the offensive, to deploy the guild of astrologers to predict Hitler's death – and feed his paranoia that his end was foretold in the stars.

The scheme was utterly batty – even on its own terms. After the war, it emerged that Hitler had no interest in astrology, and had, somewhat ironically, called it a 'swindle ... in which the Anglo-Saxons in particular have great faith'. The employment of De Wohl tells us, then, nothing about German intelligence, and a lot about British intelligence. At this time in the war, the time when Jones was beginning his work, Britain was that odd combination: contemptuous, supercilious, and desperate.

De Wohl was not an isolated case. Britain had also installed a water diviner called 'Smokey Joe' in a hotel room on the south coast, where he was instructed to use his talents to report on the status of the invasion fleet across the channel. In the words of Francis Davidson, who would become Director of Military Intelligence, it was 'a period of horoscopes, crystal-gazing and guesswork'.

The country's indulgence of such unorthodox fantasies was relatively harmless, though. Other British fantasies, whose basis lay not in mysticism but national pride and chauvinism, would prove more dangerous. The experience of the Germans in the Spanish Civil War had convinced them that at night it was near impossible for bombers to find their targets. Even by day, bad weather rendered them useless. It should not have been a surprise to the RAF that the Germans would look into blind bombing systems. Indeed, before the war Britain's own Decca Company had proposed just such a system, but the view was taken that it wasn't necessary. It was felt that British airmen were the natural heirs of Admiral Nelson and Captain Cook. They came from a proud seafaring nation, adept at navigating by the stars.

In contrast, the Germans were, in the words of Sir Hugh Dowding, commander-in-chief of Fighter Command, 'generally bad navigators'. Which made it all the more perplexing, he noted when raids began, that 'from night to night' he saw 'the tracks of accurately navigated German aircraft over the Midlands and other parts of the country'. This was doubly troubling because while Dowding's squadrons ruled the skies by day, by night they were powerless. 'The enemy enjoyed practically complete immunity by night,' he said in a meeting in Whitehall in 1940. 'Our only safeguard was that the enemy was not proving good at finding and hitting his targets.' He began to suspect that these bad navigators were indeed using some kind of trick. 'If this threat of accurate night attack were to materialize . . . we stood in the gravest peril of losing the war.'

This was the world into which Jones was recruited – a world of astrology and superstition, desperation and blithe arrogance. Jones was not a man who would be described as suffering fools gladly. Neither was he deferential. On being told just how good British navigators were at finding pinpoint targets at night in Germany, 'I was not popular for asking why, if this were true, so many of our bombers on practice flights in Britain flew into hills.'

Jones was, by all accounts, quite an annoyingly non-deferential man – in matters big and small. The official archives of his department, for instance, contain the correspondence that resulted when a hapless security officer made the mistake of asking him to tidy his desk. Like a teenager asked to tidy his room, he did not merely refuse, but recast the fug of his office as a virtue. 'You say (a) that tidiness is a principle of security,' begins his withering response, 'and (b) that because our rooms are so untidy the duty officers are never able to

check them. It may have occurred to you that there is some contra-
diction between these two statements. The reason is that you've only
grasped one half of the fundamental principle of security. You may
have heard of camouflage which is generally a better aid to security
than tidiness both here and abroad. As you are a security officer I can
perhaps let you into the secret of our policy, although I would be
grateful if you treat the information with discretion.' Those apparently scattered classified documents were, Jones assured him, nothing
but an elaborate if untidy-looking ruse. 'Our many non secret air
photographs and various documents and apparatus bogusly stamped
"secret" and "top secret" serve as camouflage with which to direct the
attention of security officers and enemy agents from any secret
papers which we have accidentally left out.' The letter, to which no
reply appears to have been received, then moved on to an extended
riff about the forthcoming introduction of booby traps to the office.

On another occasion, later in the war when he ran a small department, he was asked to fill in one of those employee assessment forms
beloved of HR departments then and now – and which are also, then
and now, loathed by the people asked to complete them. Objecting to
a section in which he had to give employees a single score for 'Personality and Force of Character', rather than just fill it in and get on
with his day, he argued that personality and force of character lay on
different axes. 'It is almost impossible to mark . . . as single assess-
ments. If I had any faith in the pro forma, I should believe that all my
staff are schizophrenic.'

He was in his own way equally scathing of the intelligence machinery into which he had been seconded. Indeed, when he started he
was little short of appalled. 'The search through the intelligence files
at once revealed the main deficiency in the Intelligence system at the
time,' Jones wrote in an official report. 'This was the absence of information.' He began to suspect, not unreasonably, that the entire
apparatus of the military was beset by incompetence, and that even
the most basic of countermeasures were not being undertaken. In
late 1939, when Britain's air chiefs were still congratulating themselves on the country's Chain Home radar, or RDF, system, he felt he
was almost alone in contemplating the very obvious corollary – that

the Germans had them too. 'Could you also tell me whether any intensive effort is being made . . . to listen for enemy RDF transmissions?' he asked in one letter, dashed off to a fellow radio expert in November 1939. 'There appears to be a danger that everyone thinks some other department is doing it. As far as I can find out, nobody else is in fact listening consistently.'

During the 'Phoney War' period, before hostilities truly began, he had to plead to be allowed to continue his work. He now had the grand, but largely Potemkin, title Assistant Director of Intelligence (Science). With his own job in doubt, it remained the case that he had no one to assist or direct within his department – that he was the totality of the department: one lone boffin in Whitehall. Initially he 'failed to get any help at all', he said, 'not even a secretary'.

In May 1940, Jones was visiting his parents in Herne Hill, in south London. There, he sat in the station and watched as the trains came in from the Channel ports, taking the British Expeditionary Force home. Train after train passed, carrying 'dishevelled troops on their way to Victoria', he recalled. With the evacuation of Dunkirk, the war had come home.

'For myself,' said Jones, 'I had no doubt regarding the importance of the work. It seemed obvious to me that while scientific intelligence could not by itself contribute more than a fraction towards winning the war, the failure of scientific intelligence to detect the development of a new hostile weapon in time might well result in national disaster. This at all times was a somewhat terrifying thought.'

With the fall of Dunkirk, there were many more such terrified people. Hitler's westward march had only one destination left: Britain. By the summer of 1940, the Luftwaffe had softened his army's path through Poland, Alsace-Lorraine, Belgium and France. It was now to be found just across the Channel. This was a time of paranoia, when the country strained its ears for the sound of church bells that would indicate the invasion had begun. Despite the mythology of a stoic nation coming together in adversity, we have a highly unusual resource that can tell us the reality.

When the war had started, the government approved an astonishingly far-sighted project, driven by a young woman called Mary Adams – described by a contemporary as a 'tiny, vivacious, brainy blonde with bright blue eyes who always dressed very elegantly'. She was in charge of the new Home Intelligence department: tasked, primarily, with assessing the nation's morale. It was her contention that before speaking to Britain, with propaganda, the government should first listen to Britain – to find out what its people truly thought.

Across the country people of 'all strata' were recruited: 'doctors, dentists, parsons, publicans, small shopkeepers, newsagents, trade union officials . . .'. They were to provide Adams with regular reports on the people they met – and what they were thinking. These were intelligence reports of a very different kind. Snatched conversations on buses, overheard gripes in the pub, mornings spent eavesdropping in cafes. What they found was often, true to today's national mythology, the dogged defiance of a nation confident in ultimate victory, that 'we shall win the last battle'. But amid that was doubt, and fear. This was not the stiff-upper-lipped resolve of a nation implacable in its belief in victory; keeping calm, carrying on, digging for victory. In Deptford, in east London, one spring report concerning 'rumours' explains, there 'has been a resuscitation of the story which seems to be worrying a good many people of a gas which paralyses the will power'. On the east coast, the same report explains, there is widespread belief that a mass evacuation to the west coast is planned. 'Armed gliders' were backing up amphibious assault troops. The upper classes showed more pessimism. The working classes, the report explained with the same eye for gradations of emotion as class, showed more doubt.

Most of all, though, as Britain's expeditionary force collapsed in France, there was a terror of fifth columnists. Britons were convinced that everywhere in their nation lay German agents – ready to take up arms when the moment came. One report spoke of the fears about a West Country clergyman who 'is stated to be instructing school children not to report parachutists'. Another report outlined suspicions of a 'house full of blind refugees who were alleged to be in possession of machine guns'. And naturally there is, this particular

Home Intelligence correspondent explains with the weariness of experience, 'the usual crop of stories about "hairy-handed nuns"'. The British had become convinced that when the parachutists came, they would be dropped as nuns in wimples. Sometimes, the rumours were so bad that – like a Stasi, but one that conducted its interrogations over tea in your front room – the Ministry of Information had to send its 'Anti-Lies Bureau' to intervene. After Dunkirk one Mrs Watson, of 145 Empire Court, Wembley, received a knock on the door from Scotland Yard – who had sent an officer to reprimand her 'for spreading the rumour that officers had fought to be evacuated before their men'.

The examples, especially the ecclesiastical ones, seem comical now, but the concerns were real. Germany had just won a spectacular victory, its armed forces clearly greatly outclassed those of Britain, and the country was obviously in line for invasion. When that invasion came, new weapons and techniques would be crucial – as, quite plausibly, would agents wielding that technology from within the country.

It was Jones's job, a job that never existed before, to find out how they would do it. In those early months, when he had access to the full range of British intelligence sources, but no idea how to use them, he began to build up a philosophy. 'I conceived scientific intelligence, with its constant vigil for new applications of science to warfare by the enemy, as the first watchdog of national defence, and to be a good watchdog, it is not sufficient to detect the approach of danger,' he wrote later. 'You must bark at the right time: not too early for the new master becomes dulled to danger by too much barking, nor too late, for he may then be overtaken by disaster.'

Doing this, he said, was not a way to make oneself popular. Particularly as, on one occasion, he formally codified the policy in official documents by explaining that giving information and deductions early 'shifts the onus of interpreting facts onto branches of the Air Staff less technically qualified to evaluate them' – or, in other words, on to people stupider than him. Sometimes, his approach meant he was accused of hoarding information, at other times of being alarmist. But timing was crucial – because it was not enough to be right; he came to realize that an intelligence service was pointless unless it could convince the right people it was right.

'Monthly summaries nos. 3 and 4' are typical of Jones's output in those early weeks. There are thirty carefully bullet-pointed descriptions of Axis technologies the existence of which he had deduced, starting with new bombsights ('An accuracy of 14 metres from 4,000 metres is claimed') and ending with 'Aerial mines in Italy'. ('An observer in Italy reports that he observed an [anti-aircraft] practice in Trieste, when parachutes were liberated from bursting shells.') As if mines suspended in mid-air and terrifyingly accurate new bombsights were not enough, in between, he warns of 'Night Camouflage' for planes, of a system for allowing bombers to maintain formation in clouds, launch catapults for aircraft, a 'Molotov's bread basket' incendiary bomb, and reports of a 'Stupefying Gas'. Introducing this shopping list of aerial horrors, Jones allows himself an aside – ostensibly an apology that the previous month he had produced no such document, but in reality a pointed dig. 'It is regretted that no separate summary could be issued . . . owing to the preoccupation of the entire staff of this branch with a specific problem.' The 'entire staff' in question was, of course, him.

As he settled into this role, he began to think that maybe it wasn't such a bad thing that he remained a department of one. Being singular had, in his view, a singular virtue: he was confident all his employees were up to the job. He was also confident that they all had the same information. 'My failure to obtain help in the early days had one interesting result, the significance of which I hardly realized at the time,' he wrote. 'An intelligence organization resembles . . . a human head. The sources of intelligence correspond to the sense organs of the head, the detailed resemblance here is in some cases remarkable, with photographic reconnaissance as the eyes and the radio listening service as the ears. The senses pass observations to the brain, where they are correlated, and a particular sound is associated with a particular visual object.'

The job is about forming connections, about remembering apparently insignificant facts and spotting the later correlations that make them significant. 'So far, no machine has been found to perform these functions nearly so well as a good human mind, and the design of intelligence organization must be such as to make it resemble a single perfect human mind as closely as possible. It follows from this that the most successful intelligence organization is likely to be that which employs the smallest number of individual minds, each with the greatest possible ability. For only then can you get these vital correlations of, say, the shadow on an air photograph with a fragment of it as coded in intercept, or with a report of a sketch from a secret agent.' In his one-man department he had, he felt, reached the intelligence singularity: the smallest number of minds with the greatest possible ability. Namely, him. It would be infuriatingly arrogant were it not for the fact that all that followed suggested he had a point.

As France collapsed and the Germans started to look beyond the Atlantic coast, Jones was, like Dowding, all too aware of Britain's terrible weakness. During the Phoney War, as the country prepared for the battle to come, members of the public would have seen searchlight emplacements and anti-aircraft batteries sprouting like angry metallic weeds around cities. While the public may well have been reassured, members of the Air Staff would have known they were almost useless. 'Our air defences, whilst excellent by day, were almost impotent by night,' wrote Jones. Britain's QF 3.7-inch AA gun looked impressive, with its barrels pointing with stern and solemn purpose into the sky. And, indeed, in a way it was. It could fire a 28-pound shell higher than Mount Everest, and then fire another one a few seconds later. The problem, though, came from that very range – it's one thing to hit a ship several miles away, quite another to hit a Junkers 88 moving at 300mph. If you pointed the anti-aircraft gun directly at the plane, by the time the shell reached it, it would be two miles away. For the gunners, the planes were near-impossible targets. What about the bombers themselves, and their own targets on the ground?

On the newly occupied soil of France, the Luftwaffe were already establishing a system of radio beacons. The beacons were a new invention in a new kind of war. Like lighthouses for the twentieth century, they flashed a clear signal. From the bearing on which that signal appeared, the navigators could very roughly deduce their position and head for port. Imagine how much more effective the beacons would be, though, if they did not stop at the French coast? Imagine if there were beacons that continued into the dark terra incognita of Britain itself – a friendly pulse from the ether amid the terrors of hostile enemy territory? What if, here in Britain, German agents were already setting up the means to broadcast signals, to draw the bombers in and guide them to their target? Those agents would have to walk among Britons, hidden in plain sight – perhaps as nuns, or perhaps as scoutmasters and stationers.

Amid all the silly rumours, there were some that were less so. Reports passed across Jones's desk of 'suspicious incidents' around West Beckham, on the Norfolk coast. German aircraft, it was claimed, frequently approached 'along a direct course for West Beckham, and then deployed inland in two principal directions'.

That was not all. A map had been found with bearings on it, pointing to nearby radio stations. It was traced to the younger son of the stationer, who said he was a scoutmaster who had used the map to practise compass readings on prominent landmarks. A plausible story? Perhaps. Except his brother, the elder son, was both an electrical engineer and, it transpired, a Blackshirt. That the only person in the area able to produce the kind of sophisticated signal necessary to draw in the planes was also the only person in the area to have actually appeared on a platform with Oswald Mosley, the pre-war founder of the British Union of Fascists, felt like too much of a coincidence.

Jones was sent with the police to investigate. He never forgot the experience of raiding the engineer's house. 'The door was opened by a patently astonished young man . . . his alarmed wife clinging to his shoulders. It is not a nice thing to ransack someone else's house, and rudely search through all the minutiae and debris of domestic life; it turns out to be so pathetically like one's own.' The police clearly needed little convincing, at one point fingering a document containing 'pages of secret calculations'. It was an old student lecture notebook. Then they came across a small wooden box, hidden and locked. They asked for the key. The engineer claimed to have never seen it, so the policeman prised it open – to find an induction coil inside, along with wire and crocodile clips. Still the engineer protested it wasn't his, so Jones read the instructions and realized, 'this was an electrical hair-remover'. It was the wife's. 'Our search had wrecked her secret – I hope that their domestic happiness survived.' The entire incident, he said, 'stands out in my memory as one of the worst things that I have ever had to do'.

And, it turned out, all the fears were entirely unfounded. The Germans had no need to place beacons inside Britain, drawing the planes in. They already had beams, outside, drawing a road in the sky.

CHAPTER 4

The Clues

'If our good fortunes hold, we might yet pull the crooked leg'

R. V. JONES

20 June 1940: One day before the meeting with Churchill

WHEN HIS CAPTORS ARRIVED, the German airman was already standing in the corner of a field, tearing up his notebook. It was June 1940 and, in the perpetual twilight of a midsummer's night, the radio operator must have known he had no chance of escape. As his parachute fluttered down, his stricken plane tumbling to destruction below, he prepared to do his duty in a less dramatic way. So it was that he did not even attempt to hide. Instead, minutes after landing, he was found calmly and methodically ripping up each page of his standard issue notebook and burying the pieces.

He was too late. Each piece was collected – hundreds of them – and rushed to intelligence officers. Over the next twenty-four hours, the muddy jigsaw puzzle that he had diligently scattered was put back together. The message the radio operator was so keen to hide was relayed to Jones the morning after.

U.K.W.	54	38	7'	N	)	Stollberg
Knicke	8	56	8'	O	)	

	51.			N			
	1	30'		O	qms	)	30 mH

Gleve	51°	47'	4	N	
	6°	6	2'	O	

	55°		N			
	2°		E	qms	)	31.5

It was the last bit of crucial information that Jones needed to understand how the Luftwaffe were going to reach their targets.

Intelligence breakthroughs rarely come from one single flash of insight. Instead, like the soiled and jagged pieces of the notebook, they are painstakingly assembled from snippets of information. So it was with the first of the German beams.

At the same time as that radio operator was enjoying his last moments of freedom, floating down above the Kent countryside, his soon-to-be-fellow prisoners of war were asleep in camps across the country. Although they didn't know it, their cells were bugged.

Jones liked German airmen. He considered them, of all the services, to be a higher class. Cleverer, more curious, more technically minded. Pleasingly for him, as the Luftwaffe's incursions into British airspace became more regular, so too did his opportunities to engage with them – with those members of the Luftwaffe unlucky enough to only make the trip one way. Primarily, he did so through reports from an RAF organization called AI.1 (K). When an engine failed, an anti-aircraft battery got a rare hit or a pilot got lost and ran out of fuel over Britain, eventually whoever survived of the crew found themselves crossing the desk of AI.1 (K). So, too, did those who did not survive.

At the start of the war, fifteen German-speaking interrogation officers were posted at RAF bases across the country, in expectation of the prisoners to come. As soon as reports of a downed aircraft came in, their role was to visit the site, which was typically already under the control of the local police. Their first job was to search the wreckage and bodies for intelligence – for useful clues that might be important to the war effort, or perhaps just helpful in teasing information out of the tired, scared and disoriented crew who, still alive, now found themselves in the local police cell. The next stage was to interview that crew – to try to establish the basic facts, and see if they might know anything of higher value. If they did, then before they were transferred to a POW camp they were sent to a very different kind of camp.

Early in the war, this camp was based at the Tower of London. By 1940, as the number of prisoners exceeded the capacity of Britain's most notorious prison, it moved to somewhere with rather fewer historical connotations: a requisitioned country house on the northern outskirts of London, which would become known as the Cockfosters Cage. Here, at a site that early in the war would become one of the most important sources of intelligence for Jones, they were treated humanely. Other moral boundaries became blurred as the war went on. Noble declarations that Britain would never bomb women and children barely survived contact with the realities of the air war. Torture, though, even in Britain's darkest hour, largely remained a red line.*

At the end of the war, Group Captain Samuel Denys Felkin, chief interrogator for the RAF and head of AI.1 (K), explained the limits he put on his staff. 'It must be remembered that an interrogator can only use words, and there's little else than his personality, determination, and experiences to assist him,' he said. 'The interrogator must be careful to avoid any suggestion of having threatened the prisoner with physical violence, immediate or future.' It was not just that it was morally and legally wrong to torture prisoners, he wrote. It was not even that doing so put our own prisoners at risk of reciprocal treatment. It was that, in his view, treating downed German airmen well was itself an interrogation tactic. 'Good conditions in camps . . .

---

* See *The London Cage*, by Helen Fry, for a counterargument.

greatly helped in undermining morale ... when German aircrew were being warned to expect ill treatment and starvation.'

Even so, quite how well some of them were treated would have probably caused a reasonable amount of disquiet in a nation struggling with rations and austerity. Felkin recounts how the best-performing prisoners would be rewarded with a trip, 'to London or elsewhere for a day's outing'. There, they were shown the sights, 'or perhaps taken to a theatre or cinema'. This was not wholly altruistic. First, said Felkin, after seeing the West End they then felt more of a sense of obligation to the interrogator. Second, as before it heightened their sense that they had been lied to. 'London was not lying in ruins as they'd been led to believe. They saw private cars on the streets, and all manner of goods displayed for sale in the shops which had long been unobtainable in their own country, and found that a plentiful meal can be obtained without coupons. They compared what they'd seen with the stories they had heard through the propaganda department of their own country. With very little persuasion by the interrogator they realized that they had been seriously misled. And they resisted interrogation no longer.'

Not every airman who baled out over Kent found himself given the full VIP treatment. As well as Jones, Felkin had many customers for the intelligence he gathered. Tailoring the right approach was, he maintained, key. 'A good interrogator is a practical psychologist with a capacity for rapid appraisal of character and an understanding of how to deal with the different types of men he encounters.' There were those who needed neither theatres nor cinemas to spill the beans – at least, according to his slightly chauvinistic interpretation of the Teutonic temperament. 'Some prisoners can be bluffed into obedience by an assumption of authority ... Germans were accustomed to automatic submission to senior officers, and to anybody who claimed authority with any show of justification. This submission was often extended to include the interrogation officer as prisoners have been told to be polite to enemy officers, and some were uncertain as to the limits of such politeness.'

For others, the weakness was not over-politeness but arrogance – or fear. 'The interrogator must remember that the prisoner is also a man who has his moments of courage, revolt and weakness. Among some prisoners are found those who are talkative and blustering, and those who are vain and with little prompting will be ready to make a show of their knowledge; among others will be found the timid people who give information for fear of punishment or to improve their own treatment. Other types are those who fall almost unconsciously into confidence, and those who are too foolish to realize that one word may complete some information previously gathered ... the interrogator must be able to recognize the intelligent from the cunning, the quiet from the hostile and the thick headed from the imbecile.'

The most powerful weapon of all, though, was knowledge: painstakingly assembled, recorded, filed and recollected. 'Small items of information and even single facts at a time were laboriously collected.' These nuggets of intelligence, from interrogations, intercepts, POW letters home and the foreign press and radio, were often superficially unimportant. Rarely would they have military relevance. And yet, in their collation and careful recording in index cards, Felkin found he could gain an edge. 'The most common method of inducing prisoners to talk was a display on the part of the interrogation of profound knowledge of all aspects of the German air force, even to the names, nicknames and idiosyncrasies of officers and their crew in a prisoner's own unit.'

'In the ideal case,' he said, 'a card might show a man's name and Christian names, his date of birth, his unit, his home address, his wife's and children's names and dates of birth (it was always impressive to be able to tell a person when his oldest son was born, and even celebrate the birthday), his dog's name and its habits, and his military career.' All of this was geared towards impressing on the prisoner the sheer futility of staying silent. 'The importance of recording such small items of seemingly unimportant information cannot be too strongly emphasized. If and when a man appeared at the centre as a prisoner, his surprise upon being confronted with such extensive information was enough to impress him with the omnipotence of British espionage, and thus to break his resistance.' After all, if they already knew so much, then what harm was there in answering a few questions?

This kind of intelligence, fragmentary and from lowly sources, was not always accorded high status in Britain's decision making. For Jones, though, it was key. He was suspicious of foreign defectors, who had a tendency to say what you wanted to hear. He was suspicious of subject experts, who he felt were too apt to believe something impossible. He was also suspicious of secret agents. 'The professional spy is usually a charlatan,' he said, before elaborating in typically florid fashion – by pointing out that the spy is the only intelligence source with 'biblical precedent'. 'You may remember that when Joshua attacked Jericho he sent in two spies to the town to obtain information ... Unfortunately for the tradition of secret services, where did they stay in Jericho, but at the house of Rahab the harlot? It is a tradition that dies hard.' Perhaps for this reason he was also especially suspicious of honey traps. 'Information which has had to jump the gap between the sexes is frequently unreliable.' The same is true 'of seduction by alcohol. Usually the statements obtained from a drunken man by an agent who is generally half-drunk himself are of precisely the quality which one would expect.'

But, time and again, the honest POW provided the final piece of information. The organization Felkin created, Jones wrote, 'reached a standard of efficiency far exceeding that in any previous war; and the flow of information which he continuously maintained ... constituted one of our most important sources of intelligence.' During the course of the war, he and Felkin built up a strong relationship, and, crucially, one in which the intelligence went both ways. Back in February 1940, Felkin had begun to pick up on a curious phrase popping up in documents and interrogations: 'X-Gerät' – or, translated literally, 'X-equipment'. He had sent round a memo, but only Jones had responded. Jones had yet to prove his worth, and had, indeed, had a request for an assistant vetoed on the grounds there was 'not enough work to justify the employment of two people'. So he had no choice but to make the trip to see Felkin in person – setting up a meeting at the Bath Club in London. At this meeting, Jones had voiced his suspicions of what the mysterious X-equipment might be. He said that it 'could only be a radio apparatus, probably an electric echo sounder', Felkin recounted. It was conceivable, Jones added – showing just how much he had already deduced – that it involved intersecting beams, 'which might possibly release bombs when an aeroplane flying along a certain beam crossed another'.

Jones needed to know more, though. He suggested that Felkin bluff the prisoners, that he let on the British 'know all about it' and already know how to jam it. As it happened, Felkin was able to go one better than that. In the National Archives, in the set of documents that now record the history of the Cockfosters Cage, there is one that feels, initially, a little out of place. It isn't a transcript of a Field Marshal's interrogation, protocols for triaging prisoners, or a set of requests from Whitehall. It is simply a receipt, from a company called RCA Photophone Limited. The job they had been contracted to perform? The installation of listening devices in all the cells.

It is not hard to withstand interrogation, if all that is required is keeping quiet. For the German prisoners who did stick to name, rank and serial number – Felkin referred to them, with the merest hint of annoyance, as the rare breed of 'intelligent' prisoner, 'cheerful, polite, and firm in his refusals' – their regular meetings with RAF officials must have seemed farcical. Each time, they were asked about sensitive intelligence and technology. Each time, the Germans refused, or lied. Never was there any sanction from their jailers more severe than stern disappointment. But while the British never resorted to torture, that did not mean they played fair. Instead, they had their own routine. It began by calling in these prisoners of interest, and asking them about Germany's military secrets. The prisoners, invariably, would lie. Then, those prisoners would return to their cells and debrief their cell mates. And Felkin would be listening.

So it is that, throughout the spring and early summer of 1940, Jones would receive transcripts of conversations between prisoners, talking about bombing Britain, about top-secret radio technology and – as often as not – about the credulousness of their idiotic interrogators. The first he saw was the one that moved Felkin to send the memo about X-Gerät. Returning from his interview, Obergefreiter 'A35' decompressed with Unteroffizier 'A29'.

'He came in again this morning about "X-Gerät", three months ago he heard something about it – a report came from France,' said A35.

'Did you tell him you knew something about it?' asked A29.

A35 did indeed know about X-Gerät. But he was a good German, and had had no intention of letting on to his interrogators. 'No,' he said, little realizing that those interrogators were listening. 'I naturally did not let him notice that I had heard about it.'

Later in the day, Unteroffizier A29 met a pilot, A34, and warned him about the line of questioning. 'That "Bombenblindwurf" business, don't say anything whatever about that!' You don't need to be a linguist to translate that compound word – blind bomb dropping.

N125, a radio operator, clearly less well briefed than the others, returned to his cell and asked A29 what the X-equipment was. 'It has something to do with dropping bombs on an invisible target,' came the reply.

Another comrade helped out the puzzled N125 with technical details, as the interrogators' gramophones dutifully etched a virgin groove, recording the conversation. This X-Gerät apparently used a 'descending current' to guide it. Yet another said he had heard rumours about its accuracy. 'There are, say, eight houses somewhere at a corner. One flies over them and drops the bombs,' he said, giving an example of how he understood it might perform.

Once the interrogators began pressing more prisoners for details, it became apparent that the equipment must be ubiquitous. Relaxing in their cells, the prisoners fed their captors continual tidbits of new information.

In this way, Jones received the transcript of one prisoner pompously explaining strategy to his cell mate, theorizing that 'blind bombing' on 'spot targets' was the future. Prisoner A212, an Oberleutnant or first lieutenant, told his cell mate that, 'it so happens I know more or less what that is. It is some sort of a codename for a directional apparatus. With it they are experimenting in dropping bombs blind. Can you imagine that? Just by the crossing of wireless beams.' The advantages, he explained, keen to demonstrate his insider knowledge, were obvious. 'It works quite independently of the human brain . . . No need for me to say: "Now I must be over London so we will drop the bombs."'

Another concluded his graphic outlining of the mechanism with the phrase 'Ho ho ho, out with the bombs.' Prisoner A57 helped out with a more credible description of its accuracy. The beam, he said, is 'no wider than one kilometre even as far as London . . . nothing complicated about it'. It became clear, though, that there was more to it, and that the airmen themselves might not have the full picture. Several times a different term was used. Not 'X-Gerät', but 'Knickebein' – crooked leg. Were they different names for the same technology?

This was far from the only time that Jones had heard this intriguing word. On 12 June, for instance, he received a 'flimsy' from 'our best sources'. The careful phrasing was a nod to those in the know that it came from Enigma intercepts. The code-breaking at Bletchley was a secret so well protected that it was rarely written down, even on top-secret documents. This intercept stated: 'Knickebein, Kleve, ist auf punkt 53 grad 24 minuten Nord und ein grad West eingerichtet.' ('Knickebein, Kleve, is located at point 53 degrees 24 minutes north and one degree west.')

The coordinates were a point in Nottinghamshire, and when others had seen the intercept, they had assumed it was the site of an illicit German beacon, used to draw in bombers. Jones had a different thought, that 'the point was the intersection of two beams, which emanated from Germany.' If so, that would also explain the word 'Kleve', a reference to a westerly German town. If you are placing a navigation beam to guide bombers accurately on to Britain, you want it to be as far west as possible. This would also fit with the evidence from a downed Heinkel III, in which a diary had been found. The sparse entries were a mixture of the banal and the momentous, sometimes on the same day.

11/3/40: Lecture on the German 'Heroic Legends'. Drank three cases of beer.

13/3/40: Forgot to wind in aerial. It tore with a nasty noise.

9/4/40: Denmark occupied

21/4/40: At Stavanger Tommy came. Three shot down

Before all this was the entry of 5 March: Two thirds of the Staffel on leave. In afternoon studied Knickebein, collapsible boats, etc.

Sadly prisoner A247, a radio operator, was no help in confirming Jones's suspicions. '[The interrogator] was only interested in finding out what Knickebein is,' he told his cell mate, on returning from another session. 'That he'll never find out.' A251, a lieutenant, responded mockingly with a reference to a children's book, *Hans Huckebein*. 'We should present him with Wilhelm Busch's book Hans Knickebein.'

But on 14 June, by which time Jones was in regular receipt of the reports, the interrogators got lucky. A new prisoner, designated A231, was transferred to the camp after being shot down over Norway. During interrogation, it emerged he was anti-war and prepared to tell them everything he knew. He stated that Knickebein was a bomb-dropping device, involving two intersecting radio beams which were picked up on one receiver in the plane. Where they intersected, the bomb was dropped – X marks the spot. Knickebein, Jones assumed, must be the name for the mysterious X-Gerät – and here was a prisoner corroborating everything he had so far deduced. But he was still confused. There were plenty of wrecked bombers. Yet there was no Knickebein equipment. Where was the receiver for this supposedly ubiquitous secret navigation device? One eavesdropping session recorded two airmen joking, after each learnt that the other had been asked about radio beams. In the kind of dramatic irony that a good scriptwriter would reject as cod, they laughed, at least according to Jones, saying, 'They'll never find it, they'll never find it!'*

The conversation persuaded Jones to ask for another sweep of their Heinkel's cockpit. Inside, the only piece of equipment that looked remotely likely as a Knickebein receiver was a common radio device. It was a landing system, a commercial device that would guide a plane in a straight line down to a runway from a few tens of

---

* This conversation is not transcribed in the National Archives, leading at least one historian to suggest that Jones – a raconteur in every sense of the word – might have indeed acted as a little bit of a scriptwriter.

kilometres away. He asked Squadron Leader Cox Walker, a radio expert at Farnborough who had previously examined it, if there was anything unusual about it. 'He said, "Wait a minute. Yes, you know, it's much more sensitive than they would need for blind landing." And of course, that was it.'

This landing device was far too well made for the job it needed to do. It would guide a plane from a lot further away indeed. Something wasn't right and, put together with the torn-up notebook, Jones now had a good idea what it was. He had enough information that, when unexpectedly summoned to Downing Street that week in June 1940, he could confidently tell Churchill about Knickebein, and the German plans to paint electromagnetic targets over Britain.

CHAPTER 5

The Chase

'The chase of Knickebein was the best fun I ever had –
but it had its frights'

R. V. JONES, *SCIENTIFIC INTELLIGENCE 1*

To UNDERSTAND HOW KNICKEBEIN works, you could do worse than listen to Prisoner A504, a major in the Luftwaffe – and an inmate of the Cockfosters Cage. In the middle of 1940, he returned to his cell having just been interrogated – and having done his duty as a good German. 'I said, "I'm very sorry, I can't say anything about it. I know nothing,"' he proudly told Prisoner A502, an Oberstleutnant (lieutenant-colonel). But given that A502 genuinely did know nothing about it, he then helpfully – at least for the listening guards – explained.

The major was a Knickebein evangelist. Knickebein, he said, 'is a wonderful invention'. 'It cuts every human weakness, every human error, because it is purely mechanical.' All a pilot had to do, he said, was listen to the sound of the Knickebein signal. 'When I'm on this beam, and have tuned this apparatus in, then I hear, "Da, da, da, da, da". And when I hear that then I know for certain that I'm on the left of the beam.' So he responds by steering to the right. 'If I hear, "Dit, dit, dit, dit", then I know I'm on the right of the beam.' And, when, instead, he hears a continuous note, 'then I know for certain that I'm on the navigation beam. That is the wonderful part of it, I need not do anything except fly my aircraft quite mechanically along this continuous note.'

What he explained in his cell that day would have been familiar to almost any halfway experienced pilot of the time. It was the set-up of a Lorenz landing beam. Lorenz beams were a commercial product, invented before the war and in use in commercial planes across the world. They were installed at some airports to guide in planes in bad weather. They were effectively electromagnetic road markings that extended the runway into the sky. If you were too far to the left, you heard a beam projecting dashes; too far to the right, you heard a different, parallel beam projecting dots. If you were approaching just right then you heard the two together. This was the clever innovation. The dots of the dotted beam were synchronized with those of the dashed beam so that both beams together played a single unbroken note. The holes in the dashes are precisely filled by the dots.

The system was designed to get round a deeply inconvenient property of electromagnetic radiation: it spreads. This spreading is an unavoidable fact of physics. Radio waves do it, microwaves do it. Even light does it. Use a torch, even a tightly focused one, and the beam width will inevitably grow to many times that of the bulb, and

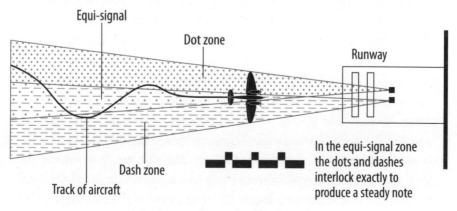

The Lorenz beam for blind landing

do so quickly. Even laser beams will noticeably widen when projected over a long enough distance. Imagine then that you created a radio beam – the equivalent of a torch but for radio. At its start, it would be narrow. It could guide a plane through fog or cloud, sending a clear signal that told the plane it was aligned and on track. But it will spread. Before it has gone very far, just as with a torch, the signal would be not a beam, but a cone sweeping a vast area. It would be useless.

Lorenz beams were a very cunning way to mitigate this, by using two beams instead. Each of those beams – one dotted, one dashed – would individually make a cone. Angle them so that those two cones just graze each other, though, and that intersection would be a tight beam. In this way, planes could be guided in to land from as far as 20 kilometres away. But not much further, it was thought. Britain, at least, had not even considered the possibility of using it for anything other than medium-range runway approach.

After hearing the taunts of the prisoners, Jones had reconsidered the electronics of their wrecked Heinkel bomber. In the smashed cockpit was its Lorenz receiver. This was not in itself notable. Most German bombers had them, and they were treated as largely unremarkable. What was different, though, was the engineering. Speaking to Cox he learnt the engineering was vastly more sensitive than it needed to be. It was, in fact, sensitive enough that it could pick up Lorenz beams from over 200 kilometres away.

The beams system, which the Germans called Knickebein, and whose presence Jones had long suspected, had been hiding in plain sight. And if the wreckage of the Heinkel told him the 'how', the torn-up paper from the downed airman told him the 'where' – it told him the locations in which the beams were based. The second location mentioned was Kleve – Jones guessed that what had been transcribed as 'Gleve' from the muddy fragments stuck together was actually the town in western Germany described in the intercept, sometimes known in English as Cleves. The first was given as Stollberg. The most obvious Stollberg is a town in the far south-east of Germany, exactly the opposite side of the country from where you would want a beam. Stollberg Hill, however, is on the border with

Denmark. And a hill, from where a tower can project further over the horizon, is exactly where you would want a beam.

After the climactic meeting with Churchill, it had been agreed that flights would at last go up, to seek out these beams. Yet even so, as an ageing Anson plane juddered and shuddered into the June sky, Jones was nervous. Just hours earlier he, a 28-year-old junior scientific officer, had talked over the heads of some of the most important men in Britain – to persuade the *most* important man in Britain that he, Reginald Jones, was right.

His evidence that the beam existed came from intercepts, defectors, prisoners and downed planes. His theoretical argument – that the approach the Germans were taking was possible at all – came, though, from radio experts. It came from, in particular, one radio expert: Thomas Eckersley, a physicist from the Marconi company who had been seconded to the UK's listening service. He had shown in a scientific paper that a beam would follow the curvature of the Earth, bending slightly towards the surface and travelling further still beyond the horizon. The reason for this is complex. It's not because of gravity; that effect is tiny. Partly, it can be explained by diffraction, the phenomenon whereby when a wave encounters an obstacle – in this case the ground – it bends around it. Partly it is because the atmosphere is denser closer to the ground, and the beam is refracted – bent like light going through a prism.

But the consequences of this, for the German radio scientists, were profound. If the paper's conclusion wasn't true, if radio beams don't curve at all, then to follow one you would have needed line of sight. Since the Earth is round, to stay in sight of a beam starting in Kleve, you would have to climb higher and higher until, fairly rapidly, you would be higher than a plane can fly. Even Stollberg Hill would only take that beam slightly further. The distance from Kleve to Nottingham, for instance, is 317 miles. To fly high enough to have line of sight of Kleve above Nottingham, you would need to be at 60,000 feet – twice the height of Everest and three times the cruising altitude of a Heinkel He 111. If, however, the beam bends so that the bottom of it curves with the Earth, then a plane could travel further before it hits its ceiling – far enough to reach the great industrial centres of the north of England

Eckersley's calculations showed that on the frequency used by Knickebein, the beam would still be able to be found at a usable altitude above much of Britain. It was these calculations that Jones had shown his mentor Lindemann, who had in turn been convinced that the theory was credible – credible enough that he arranged for Jones to be invited to Downing Street. And there, relying in part on these calculations, Jones convinced Churchill that the beams were real, were a threat – and they should find a plane to search for them. Later, captured documents showed that the Germans anticipated that for a bomber at 20,000 feet the reliable effective range of the system would be 270 miles – a calculation that assumed a modest amount of bending and, as it turned out, less than occurred in the real world.

There was a problem, though. After the meeting, Eckersley changed his mind. 'Mr Eckersley gave us his opinion that the useful limit of the Knickebein beam at present . . . does not extend to any great distances within our coastline,' wrote a somewhat incredulous Jones. Any signals that were found would be 'in the nature of freaks', due, perhaps, to unusual weather. 'The situation,' said Jones, 'was not without humour . . .' He had just convinced the Prime Minister of Britain to personally authorize flights to find these beams, which his best radio expert now said could not be found. 'Mr. T. L. Eckersley, who is a world authority, had stated that the radiation just would not bend around the earth's surface.' Yet at that very moment the over-stretched RAF was, largely at his behest, sending up a plane to find these impossible beams. His feelings, Jones later said, 'can be imagined more readily than they can be described'.

The significance of this difference of opinion is hard to understand now. Kleve was chosen for the beam because it was the western-most part of Germany. Since that transmitter had been built, Dunkirk had happened, France had fallen and the writ of the Nazis extended considerably further west. Beams can be moved. It didn't actually matter whether Eckersley was right that they could bend, as he suggested in his earlier paper, or right that they couldn't bend, as he confusingly suggested in his retraction of it. Nottingham might not have line of sight with Kleve – but it was in line of sight with Calais. Even so, on that night in question, a decrepit Anson plane – which was all the RAF could spare – headed off into the dark of the Midlands. Jones felt that his reputation rested on what it found.

The pilot was Flight Lieutenant Bufton. The observer was Corporal Mackie. They were there to find Jones's beams, but neither knew the purpose of their flight. They had been simply told to 'search for transmissions with Lorenz characteristics', and they had been told to do so in the vicinity of Derby. With its importance for aircraft manufacturing, Jones had an idea it might be a target. Even if it wasn't, there was a good chance another city in the region would be. During its northwards flight from Cambridgeshire, the plane would intersect any beam targeting one of the great Midlands industrial centres. All the same, so much was left to luck. The beams that Jones hoped to find could have been anywhere over Britain that night. They could – from a strategic point of view, in fact, *should* – have been switched off at that point. It would have made a lot more sense if they were only turned on when the bombers were ready to fly.

The sky was dark and cloudy and, at first, silent. But then, south of Spalding a series of faint dots were picked up. As the plane shuddered onwards, they grew in strength until they became a continuous note, lasting just a few seconds. Finally, as that note receded, the ageing British bomber flew into not dots but dashes – until they too were lost. The plane returned once more to radio silence and an oblivious Lincolnshire sky. They had found the 'Lorenz characteristics' they were looking for. It was, said Bufton, a 'narrow beam, approximately 400 to 500 yards wide'.

Below, the people of Spalding slept. Above, the Anson had crossed an electromagnetic road, laid in the sky, to take hundreds of bombers into the heart of Britain.

But his flight was not done yet. Continuing north, the silence of the June night was broken once again, this time by dashes. Over Beeston, they found another beam, a mirror image of the first. It was the cross beam.

An aircraft approaching a runway needs one set of beams to guide it along the correct approach, as its final destination – the ground – tends to be pretty obvious. So the commercial Lorenz system needed no second beam. A bomber, however, does. The first takes the pilot to his target. The second – cutting across – tells the crew when to release the bombs. All a Luftwaffe pilot had to do, as Prisoner A504 explained, was fly along one of the beams, then listen out on a different frequency for the second one. When it arrived, when its dots became a continuous sound, he opened his bomb bay.

They did not find this point, where the beams intersected – not directly. Instead, the Anson flew through each beam in turn, for each finding a single coordinate lying on their path. But that was enough. Jones and colleagues knew the beams started in Kleve and Stollberg Hill, they knew they went through these other coordinates. From that, they could draw two lines on a map, the route of the beams that night. Continue the lines, and you get the crossing point: the city of Derby, containing the factory where Rolls-Royce built the engines that powered every Hurricane and Spitfire. X marked the spot.

Jones did not know it, but at around this time across the Channel Goering had outlined his strategy for defeating the RAF. It relied, he had told his subordinates, on 'nuisance raids' on 'aeroengine factories' – factories like those in Derby.

Until now, Knickebein had existed, for Jones, as an idea – slowly taking form in his mind. Suddenly, it was all too real and all too

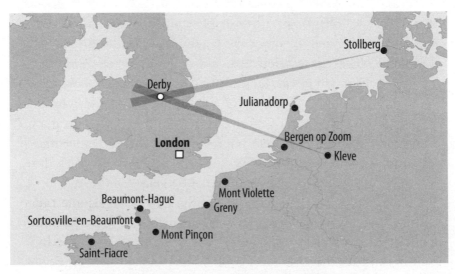

Knickebein beams paint a cross over their target

terrifying. But the deductions were obvious. 'The excellence of these observations, when it is remembered that they were taken in heavy clouds and darkness and the pilot was unaware of the existence of the beam, will be appreciated,' he said. 'It appears experimentally that the accuracy of this method of bombing might be as good as 400 yards or even less.' In other words, this was not quite precision bombing as we would know it today, but if you got a massed bomber force to the cross beam then you could be pretty sure the target would be hit. 'As there is now evidence that the beams can be swung over considerable angle, the danger of the threat is obvious.'

Now, the question was, what were the RAF going to do about it? The answer, initially at least, was to install garden sheds on 300-foot-high radar masts.

'Until you've tried it, you have no idea of the effort required to haul a bucket up 300 feet,' Harry Spencer later recalled. Spencer, better known as Spenny and even better known by his call sign G6NA, was in his mid-thirties when war was declared. He had been interested in radio since he was a child, a member of the amateur radio community, who tinkered and fiddled and chatted to each other – enjoying the thrill of existing in a private community, layered on top of the unknowing world around them. It seemed natural to offer his expertise to the country. He just didn't expect to be applying it like some kind of modern stylite, sitting in a box on a pole, lugging up water for the kettle.

When Knickebein was detected, he was given the job of monitoring it. But his mission was, he said, 'so hush-hush that we were not told our destination'. Taken by a train to Newcastle, then driven by a WAAF to Usworth Fighter Command Station, Spencer felt the glamorous nature of the top-secret undertaking was slightly undermined when for the final leg he was told to hitch a lift on a brewer's dray, making its deliveries to a Northumberland village.

This last stop was the site of one of the new Chain Home radar stations. Starting in 1938, great pylons began appearing on the south and east coast. To the local residents, they were mysterious, a little exciting, and just the sort of thing that propaganda posters urged you not to talk about.

To today's radar experts also, they look a little strange. These are not the rotating dishes we see on modern control towers. It was the more compact and versatile approaches taken by other countries that would go on to provide the post-war radar network. But, in those critical years when Britain looked out on the darkness of Nazi Europe, these were the country's eyes. And, if the local residents had looked again at these radar stations in late June 1940, they would have appeared even more mysterious. Because on the top of them was Spenny, in a box. The box was a tiny garden shed ('with floor,' Spenny feels moved to add – lest we feel he was treated poorly) 'lashed to the top of a 300-foot mast'.

'It was very cold up there,' he said. But he got to work, jury-rigging receiver equipment so that if the Knickebein swept this way and the conditions allowed traces of it to come down near ground level, he could spot it. Home comforts were, he found, sparse. 'After a week or two an RAF corporal was attached to me so he could take over the work. Feeling cold at night (even though he had recently come from Narvik) he soon scrounged a one-bar electric fire and a mile drum of signal cable to make enough parallel lines to transmit useful power up the tower. He then "found" a Swan kettle, a coil of rope and a bucket.' As the strain of hauling it became too much, the corporal – whose resourcefulness Spenny clearly admired – in his words, 'came by' a pulley and some scaffolding. 'From the top of the tower no habitation was visible for miles, so how he found all the bits and pieces remains a mystery.'

It was the same across the country. The military was begging, borrowing and (though Spenny never accused the corporal outright) stealing their way to a functioning radio listening system. Further south Cyril Banthorpe, another radio enthusiast, was ordered to Whitehall. 'I could not think I was important enough,' he later recalled. But he was. Robert Watson-Watt, the father of British radar, wanted to see him. 'I was ushered into the great man's office. Sir Robert came straight to the point. "We are sending you to Felixstowe to listen for enemy signals, using an aerial attached to a barrage balloon. A receiver will be waiting for you".' The balloon, though, a vast construction over sixty feet long that could be floated to 15,000 feet, would not. So

Banthorpe asked the obvious question. 'How should I get hold of a balloon, sir?' To which the reply came, 'Just pick one up on the way.'

'And so I set off. Between Ilford and Romford I passed a barrage station, and drove up to the gate where the armed sentry demanded to know my business. I said, "I need a balloon." The sentry did not seem to find this at all odd.'

So it was that with sheds on poles and receivers on balloons, as well as more flights, the Y-service staff maintained a continuous watch – listening out for the sweeping Knickebein beams, listening to see where Goering would turn his gaze for his next target. What would be done with the intelligence once it was found was another matter. When prisoners of war realized Knickebein was no longer secret, and were questioned about it, they were heard to discuss the ramifications. 'Can they make any use of the fact they know about Knickebein?' asked prisoner A510. 'No,' replied prisoner A538, 'they can't disturb it. The most they can do is to listen in to it and turn the A.A. guns on to it – or send up the fighters.'

A538's assessment was very nearly correct. Knickebein transmitted on around 30 MHz, meaning the beam oscillated 30 million times a second. The first stage to countering a beam is to reproduce it, or something like it. Embarrassingly for Britain, the country appeared to have no equipment that could match that frequency – mimicking, or masking, the beam – at sufficient power.

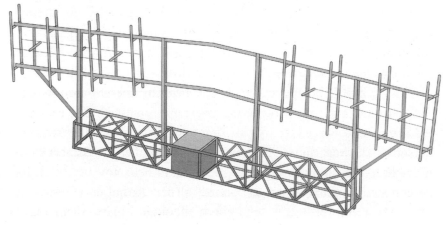

Knickebein beam apparatus

There is an elegant method of dealing with a beam like this. Given time and ingenuity, it is possible to spoof the signal, to fill the air with dots or dashes at the right intervals and right frequency – so that the beam is bent, the aerial symphony played out by Knickebein is imperceptibly interfered with, and the planes fly on to a different target. There is also an inelegant method: fill the airwaves with so much noise at that frequency that, like a vuvuzela blasting over the top of a Chopin Waltz, the intended signal is completely obscured. The truth was, in those desperate months there was no time for elegant methods – no time to actually bend the beam. The Blitz started in earnest in September 1940, and it was a time to make do and mend. The RAF needed something, anything, that transmitted on the same frequency as Knickebein and was powerful enough to interfere with it. But how? No radio sets that Britain then owned fitted that description.

Then, someone remembered diathermy machines. Diathermy machines are not used for broadcasting, they are used for cauterizing after surgery. They generate radio waves powerful enough to melt flesh, and they do so, by chance, on the same wavelength as a Knickebein. Two diathermy sets were quickly requisitioned from hospitals and put in vehicles, one driven to Boscombe Down in Wiltshire, the other to Wyton in Huntingdon – ready to pounce. Then the RAF sent up planes to listen to them, as they attempted to jam the beams. They were, they found, woefully inadequate, with neither the required range nor the jamming effect. So, at least according to the official history of Britain's radio countermeasures, 'More diathermy sets of different patterns were therefore obtained from hospitals, and 12 of the most suitable were selected and modified.'

This form of language, in particular the 'were therefore obtained', did not quite tell the whole story. Later Air Vice-Marshal Edward Addison, who was in charge of organizing the countermeasures, explained the 'technique'. He enlisted the help of a doctor, and gave him some very specific orders. 'Go to Moss Bros and hire a flight lieutenant's uniform, then go around all the hospitals where you know there are diathermy sets, pinch them, and bring them to us here and we will modify them.' Anyone unlucky enough to end up in

Britain's wartime hospitals would have to do without cauterization equipment. On the other hand, Addison could reasonably argue, if his plans worked as hoped, fewer people would be in a position to need it.

In late June and early July these sets were installed in twelve carefully sited police stations along the south and east coast, to form a 'jamming screen'. For now, Britain had the rudiments of protection from Knickebein. In those ad hoc jammers-cum-fleshmelters lay the nation's first systematic radio countermeasures. But it was clear this was a stopgap. The country needed more than diathermy sets. It needed more than disguises from Moss Bros and radio enthusiasts standing on towers.

In early August, Air Chief Marshal Hugh Dowding called a meeting. Addison was there, as was Eckersley. Jones, on whom so much of the response had depended, was not. At this meeting it was still not clear how important Knickebein was – in part because, it transpired, it was yet to be fully utilized by the Luftwaffe. Just two months earlier Winston Churchill had talked of the 'lights of perverted science', shone from Nazi Germany, that threatened to sink the world into a new Dark Age. In this radio war, in this use of light that cannot be seen, was the literal manifestation of Churchill's metaphor. In response, Dowding said at that meeting, 'we should put ourselves in a position where we can immediately jam the ether'.

New organizations were needed, for a new kind of war. Scientists were needed to create the devices that could counter German radio weapons – they would come together in an organization called the Telecommunications Research Establishment – and to use such devices in the field another organization called 80 Wing was set up. Its job was to jam, deceive and distort the German radio weaponry. Or, in the words of its motto, to bring 'Confusion to Our Enemies'.

The creation of 80 Wing was a sign to Jones that his concerns were being taken seriously. Knickebein, he felt, was almost defeated before its use had begun. But, even so, he had a persistent niggle. While Knickebein was indeed a blind bombing system and, until it was discovered, a very good one, it was definitely not the blind bombing system described in the Oslo Report.

CHAPTER 6

Cornish Patsy

'Confusion to Our Enemies'

MOTTO OF 80 WING

T HE SKIES ABOVE CORNWALL could be a lonely place. Gunther
Dolenga, a German bomber pilot, was hunting for ships at the
western approaches of the English Channel. He had hoped to mark
this, his fiftieth mission, with a kill, but he had failed even to find a
target. Plotting the plane's position from snatches of coastline, or the
white breakers that showed through the gloom, he and the other four
crew of his Dornier 217 bomber had patrolled without success. They
were looking for easy merchant navy shipping to attack; they found
nothing. Finally, frustrated, at around two in the morning, they
decided to call it a night. That was when their problems began.

The blackout in war-hardened Britain was by now absolute. Without a clear moon in the sky, it had been difficult for the German crew
to match the southern coastline of Cornwall to what was shown on
their charts. They would find an inlet or a rocky outcrop and it never
quite tallied with what they were expecting to see. As they scouted
along the edge of Britain, they eventually gave up on precise navigation and decided just to fly south. From there, things should be easier.

The beams in theory allowed precise navigation inland. But other, simpler, radio systems helped planes once they returned to the safety of occupied Europe. Dotted along the Atlantic coast was the system of German beacons, electromagnetic lighthouses that they had installed after invading France. Each, just like a lighthouse, transmitted its own unique call sign, so that German planes knew which one they were looking at, and, vitally, they could also work out where they were in relation to it.

Unlike lighthouse beams, the beacon beams did not rotate. Instead, in a sense it was the navigator using them that did. In the German planes there was a receiver that turned. As it did, there was a regular pattern. The signal from the beacon was strongest when the receiver was perfectly aligned, then diminished to nothing as the receiver rotated away from the beacon, before growing again. This wave-like pattern gave the navigator a rough bearing to the beacon. If there were two beacons, he had two bearings – and could fix his position at the place they intersected.

For the Dornier plane flying that night over Cornwall this technique seemed to be working. As the radio operator scanned the air waves, he tuned into the signal from the beacons at Paimpol in Brittany and Tocqueville in Normandy. At last they had a position, and the maps did not matter. With the signal growing stronger, so too did the relief of Dolenga, the pilot. He had thought, before hearing the welcoming radio pulses from France, that he really was lost. Now he knew there would at least be enough fuel to get to a friendly airfield, and probably home as well. There, alone in the endless black sky, they relaxed as the twin engines of the Dornier droned comfortingly, taking them to – they thought – safety.

The Dornier continued until it was directly above the Paimpol beacon, its signal pinging loud through their receiver, and then, at 4 a.m., Dolenga set a bearing for their airfield at Evreux. They would be home for breakfast, and for the bottle of sparkling wine he had saved to celebrate his half-century. Except, Dolenga still had a nagging feeling – that he tried to suppress – that nothing was quite as it should have been. As he continued on a trajectory that should have taken him over familiar territory, he realized he could see nothing he recognized. Worse, the beacons were not behaving as they should. Paimpol was fine and clear, but the others just weren't where the charts said they should be.

Resigned, and running out of fuel, he descended for an emergency landing on the first suitable field. Something had gone dramatically wrong. He suspected he had been blown off course and crossed into France farther to the north. As the plane cut its throttle the ground rushed up faster and faster, until with a jolt its wheels hit the field – churning up the damp mud. In its last few feet the plane overshot and flopped into a ditch, its back breaking. Dolenga thought his humiliation was complete. He, an experienced pilot, had got lost and destroyed the Dornier. Now he was going to have to find the local Wehrmacht, arrange for its retrieval and beg a lift back for an embarrassing breakfast in the officers' mess.

But as it happened his night was going to get a lot worse. Because what he had thought was the Cornish coast was actually the Welsh coast. When he believed he was crossing the English Channel it was actually the Bristol Channel. And the comforting signals from the

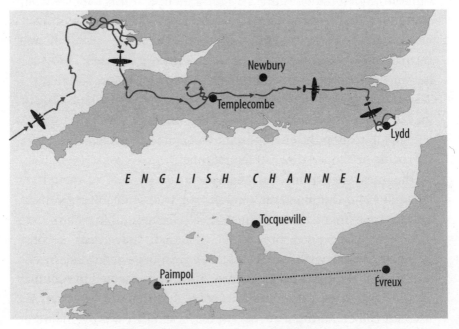

The ill-fated flight of Gunther Dolenga

German beacons? They were a British deception. When the German beacons switched on, radios in Britain picked up the frequency, amplified it, and retransmitted it – in the case of Paimpol from a site in Templecombe in Somerset.

From out of the dark gloom the Dornier's crew made out an approaching soldier, with the distinctive silhouette of a British tin hat. He had come to welcome them to the south of England.

80 Wing had just claimed another scalp. Dolenga never did have his sparkling wine.

In August 1940, Sir Hugh Dowding, commander-in-chief of Fighter Command, had called for scientists to jam the ether. It was to do this job that 80 Wing was created – or rather re-formed. The unit had briefly existed in the First World War to coordinate fighter squadrons. Now Air Vice Marshal Edward Addison, who had so impressed the top brass with his grand hospital diathermy heist, was tasked with leading it. His job, in this modern war, was to coordinate the new electromagnetic squadrons.

From Aldenham Lodge, a country house in Hertfordshire, 80 Wing took on the role of radio countermeasures. Jones's job meanwhile was to work out what the Germans were doing – to sift the vast river of intelligence through the sieve of his mind, until all that was left were the few nuggets of gold that indicated what they were up to. There was no point in knowing what the enemy planned, though, unless he and his team could then do something about it. It was 80 Wing's job to stop the enemy – part of an entire organizational structure being developed downstream of Jones.

The 'meacons', from 'masking beacons', were one of their most ingenious and durable strategies. At worst these decoy radio lighthouses, indistinguishable from the real German radio lighthouses, made German planes forever question their judgement, second-guessing their navigation. At best, the fake ones acted like Sirens, luring pilots on to the rocks. As new masts were erected in occupied Europe, and new techniques tried out, the meacon was often the simplest riposte. If there are two signals, each the same as the other, which do you choose?

The meacon was not, though, how they chose to deal with Knicke-bein. During the summer, as more towers began construction along the French coastline, Knickebein acquired the British codename 'Headache'. Among the group of boffins at the Telecommunications Research Establishment, who from their Swanage headquarters supplied 80 Wing with their apparatus, there was a race to produce something that worked better than repurposed hospital equipment to alleviate this national headache – in truth a potentially terminal migraine.

They called these countermeasures, naturally enough, Aspirin. Later in the war rumours, which the RAF saw no reason to dispel, suggested that through Aspirin the beams had been bent. When bombs fell on Dublin by mistake, it was blamed by the Irish on the work of the perfidious English oppressors, mischievously redirecting the Germans to attack the former province. When a bomb fell on Windsor Castle, there was a call from a furious groundsman asking the signals team to be a bit more careful in future. Obviously it was the RAF's patriotic duty to divert the Knickebein beams, but in doing so they didn't want to inadvertently blow up the King's prized hydrangeas.

The Germans, too, heard these rumours. After the war, Johannes Plendl, the German chief radio-navigation scientist, confidently told interrogators that pilots had been directed in semi-circular arcs away from their targets, uselessly dropping their bombs elsewhere. General Wolfgang Martini, director general of German Air Force signals, said the same. His post-war interrogation report states: 'The crews using Knickebein soon reported that the beam was being diverted, and that British fighters were being vectored on to it. Several weeks were required to prove that the beam was really being diverted. After some weeks, experienced signals officers were sent out with the bombers and reported that countermeasures had in effect been taken by the British.'

Countermeasures had indeed been taken, but they weren't quite as sophisticated as the general believed. They weren't bending beams. The Aspirin equipment was not purpose-built countermeasures, but

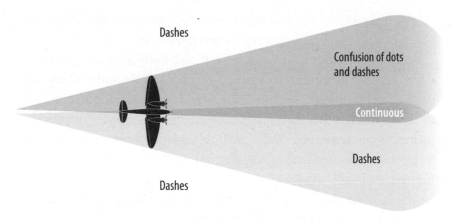

Dashes

Confusion of dots
and dashes

Continuous

Dashes

Dashes

Aspirin countermeasures scrambled the beam

just modified army transmitters. These were tuned to the Knickebein frequency, where they transmitted identical dashes to those used by the German navigators. At twenty-eight sites around the country, they effectively broadcast one half of the Knickebein signal in all directions. So instead of a clear cone guiding bombers on to the beam, the entire sky was filled with a continuous stream of dashes precisely the same as those coming from occupied Europe.

Unlike the Knickebein signal, they were not directional. Almost everywhere the pilots flew, they heard a beam apparently directing them to turn left. The only place they did not hear this precisely was if they were in the genuine beam, where there would be a dot overlaid with a dash.

If the diathermy sets made it hard to follow the route, these new transmitters meant that everywhere the planes looked there was another route. The once-clear electromagnetic road in the sky had become a spaghetti junction of potential paths, with no signpost pointing to the correct one. What they did not do, though, was truly bend the beams. To do that, 80 Wing would have had to precisely synchronize dots and dashes with those broadcast by the Germans, to shift the equisignal – the place where a continuous note could be heard. This was perfectly possible, and work to do so was begun. Robert Cockburn, the scientist chosen to head the Telecommunications

Research Establishment, lamented that the 'exigencies' of the conflict meant he and his team never had time to try it out.

Such a system, he said, would have had clear advantages. 'Even when the enemy begins to suspect trouble, controversy is always likely to arise between the sponsors of the system and their critics on its intrinsic reliability.' Such is human nature, that those invested in the beams' success would have inevitably sought to defend it, so long as the existence of jamming was ambiguous. This would have meant that the Luftwaffe would have ended up using a useless system for longer. 'Ultimately, the introduction of subtlety into any counter-measure is an attempt to gain time.'

But it was not to be. 'The increasing tempo of the enemy's attacks demanded the most immediate application of countermeasures, however obvious . . . By this time, the enemy was systematically blitz-ing out industrial towns, and it became essential to set up a carpet of jamming as quickly as possible with a minimum of elaboration.'

On the night of 24 September, midway through a raid two Knicke-bein beams swapped frequencies. It was the first sign, for British signals intelligence, that Aspirin had been rumbled. The move was, an official report stated, 'unquestionably an attempt to obtain relief from our countermeasures'. The German radio operators also took to switching the beams on only when they needed to be used – a very basic countermeasure that, mystifyingly, had not been undertaken from the start.

The only surprise was that the move itself had taken the Germans so long. For weeks, prisoners of war had been heard grumbling about the British beams, which had now been constructed in eleven sites along the coast of Britain. As early as the start of September one air-man said that he knew there were countermeasures, and that they were so effective they 'obliterate the weak beams and turn aside the strong beams from the targets'.

And yet their commanding officers had not acted on this. They continued planning and executing missions as if the beams still worked. Later, Jones received intelligence that he believed provided the only plausible answer for this delay: no one was prepared to inform Goering. 'It seems the Reichsmarschall had such faith in the system that he declared it to be infallible. Such was totalitarian discipline that it was not until a month after Cherbourg and Dieppe beams had been found to be effectively jammed that his staff dared to tell him the truth.' This very much appealed to Jones's sense of the absurd. 'Presumably, during the intervening period, his Geschwader [Luftwaffe squadron], like the subjects who could not see their king's invisible and imaginary robe, made their nightly sorties to bomb on imaginary beams.'

The downed airmen, it has to be said, were less amused. In the first week of September, the view was already developing that the beams, even when they did work, were a trap – that since the diabolical British knew about these electromagnetic highways, they were waiting in ambush like highwaymen of the sky. 'The searchlights are close beside the guiding beam and illuminate everything,' moaned Prisoner A565. 'They don't shine into the beam, the devils, always next to it.' On the beam itself, meanwhile, the nightfighters picked off their prey at will. 'They patrol ... continually.' Prisoner A602 agreed. 'We've lost a number of crews owing to that damned "Knicke". They get to the right and left and riddle the aircraft.'

It was only gradually that Jones and his colleagues realized the true power of their countermeasures. Even when they were only marginally effective – in the technical sense – they had great psychological value. Simply by convincing the Luftwaffe that the RAF knew their plans, and had their own counterplans, they were able to sow fear and mistrust.

The early years of the war have become, in British mythology, the Finest Hour – but also the Darkest Hour. Alone, the country stood up to a war machine the likes of which had never been seen. A highly trained, highly efficient and highly technological army threatened to snuff out the freedoms that had lasted a thousand years. In the towns and cities, the villages and countryside – yes, even the beaches and landing grounds – Britons waited night after night for the ringing of church bells that would announce an invasion.

All of this is true. The Nazis were indeed highly trained, highly efficient, highly technological and a lot more besides. As a military force, they were indeed also planning on bringing Britain's freedom to an end. But this understandable picture misses something else. It both elevates and dehumanizes a foe that was very human. This was as true for Jones and his colleagues as for the general public. 'Intelligence has a natural habit of viewing the enemy as his protégé, and so to extol his potentialities,' he said. Because he spent his time anticipating and uncovering German technology, and then persuading his superiors to take action, he was in an odd sense the advocate and champion of his opposite numbers.

Yet Germans too made mistakes. They miscalculated. They missed things. They too overestimated their enemy. None more so, perhaps, than the young and frightened men sent into the dark and unforgiving skies of Britain. One report at the time outlined the circumstances of the crash of a Heinkel 111, showing just how frightening it could be alone in the clouds above a hostile Britain. 'This particular aircraft had, as an individual target, an aircraft factory to the north of Birmingham ... They picked up the beam (at 13,000 feet) over the Channel and started to fly along it. After a time the [signal] became variable and then disappeared altogether, and no matter how they tried they could not pick it up again.' They began to worry.

For German pilots, British skies were a menacing place. Flying alone in the clouds, they squinted into the blackness for the shape of a plane that would mean death – as with all bombers, the truest defence against a nightfighter was to never meet the nightfighter. It was hard to feel part of an all-conquering military. When you are only partly in control of your fate, you resort to superstition and myth. Rumours that the beams were being bent by the British had spread through the Luftwaffe, who feared not that they were being covertly diverted to bomb Dublin, but instead that they were being directed into the guns of the waiting RAF fighter squadrons.

The report of this one doomed Heinkel continued: 'Panic then seems to have overtaken the crew of the aircraft.' They began to imagine the equipment was being interfered with too, that the artificial horizon – used to show the plane's orientation – had stopped working. The wizards in the RAF's electronic countermeasures department were, they thought, using mysterious science to bring them to their doom. The pilot began to lose control, only confirming the crew's hypothesis that their equipment was being nobbled. 'Eventually the observer jettisoned the bombs and then he and the [radio] operator baled out. The pilot and gunner were killed when the aircraft crashed.'

Aileen Clayton, a British radio intelligence officer, was one of those who remembered this time with guilty pleasure. Or, as she put it, 'unholy glee at the alarm and despondency of the Luftwaffe crews'.

Yet, still, only hints of this were trickling through to the German high command. As the official British history of the Luftwaffe put it after the war, 'Such was the totalitarian discipline . . . that no one yet dared admit the failure of Knickebein.'

Slowly, as the RAF regained physical air supremacy over Britain, 80 Wing gained electromagnetic supremacy. Equally importantly, the country regained some sort of psychological supremacy. But, throughout, Jones kept worrying.

Right at the start of the Blitz, two POWs were recorded in conversation. A395, an Oberleutnant, had just been asked at his interrogation about Knickebein. 'Now he wants to know what Knickebein is,' he said to A427, a Leutnant. 'He said that our Geschwader needed it when the clouds started at the end of September. I told him I did not know anything.' A427 was disgusted that the secret was out. 'It's enough to make you sick. Someone must have blabbed.' A395 agreed, but said they should keep it in perspective. 'Knickebein is not so important. There are other things besides Knickebein.'

Here was another hint that Knickebein was not the Luftwaffe's best navigational tool, not the tool mentioned in the Oslo Report. Jones was now increasingly worried that it was merely its first tool – but, fearful of being, in his words, the watchdog that barked too soon, he needed to find out.

Just before war broke out, when he was based in Teddington, Jones had been part of a group of physicists dispatched, along with all good citizens at the time, to dig slit trenches, in case of air raids. The location they were assigned was a nearby field. Unfortunately for them, that field was also a hockey pitch – and the hockey team did not take kindly to the idea. Jones recalled how an 'irate little woman' called Vera Cain drove them off the pitch. She was very intimidating. She was also, for Jones, alluring. In March 1940, they married. They started their married life in a top-floor flat on Richmond Hill in south-west London. Every day, Jones would leave her here, with her views over Kew Gardens, and travel the nine miles into central London. As the Blitz began to disrupt trains, increasingly he did so by car.

Anyone following this tall young man on his commute would have then seen him park his car outside number 54 Broadway, an imposing sandstone building that lay almost equidistant between Buckingham Palace and the Palace of Westminster. They might have been mildly puzzled as he walked in, past a plaque bearing the details of a company that did not obviously merit such a salubrious and well-situated residence: 'Minimax Fire Extinguisher Company'. Such a building in such a location surely at the very least merited the tenancy of a small embassy, or perhaps the fisheries department. On the other hand, they may not have been puzzled at all. It did not do well to investigate too closely what went on in these streets.

Throughout the war and for two decades after, this was the headquarters of the Secret Intelligence Service – these days better known as MI6. Here, Britain's spooks countered foreign threats. And, in one little corner of it, they made room for a desk and the country's smallest intelligence operation – Reginald Victor Jones.

One morning in September 1940, unusually, he did not make the commute to his desk. Instead, he went to a country house north of London. The night before, he had received a phone call from Bletchley Park, where the German Enigma code was now being regularly cracked. 'We've got something,' shouted one of the intelligence men. 'God knows what it is, but I'm sure it's for you. Can you come down in the morning?'

A decrypt had come through, detailing the construction of high-powered beam transmitters near Calais. There was also evidence from prisoner interrogations that these transmitters were somehow associated with the Heinkels of K.GR. 100, an elite German bombing unit that specialized in navigation. The most troubling details of the report were the coordinates given of the transmitters, in the Enigma intercept. It was not their location, but their precision. The position of where they were to be built was being described to the nearest foot. 'To measure positions to this accuracy is a tedious process; astronomical methods are known to have been employed,' wrote Jones, in a rushed report. 'The Germans, who are probably in a hurry, will not have wasted time in unnecessary overdetermination of site positions, unless their bombing method merited this accuracy.' If your beams only have an accuracy of a mile over their target, then caring about whether they are a foot either way is obviously completely pointless – the extra inaccuracy from how you site the beam is then completely irrelevant. What worried Jones was what it implied if it wasn't pointless – if the accuracy mattered. 'This accuracy may,' he added, 'at first sight appear incredible.' Indeed it did.

He did a back-of-the-envelope calculation. If, as he now believed, the new beams were an automated bombing system for pathfinder pilots, then they would give a bomber accuracy over London of between ten and twenty yards. British bombers at the start of the war were lucky to drop their bombs in the right town. Even with Knickebein working properly, German bombers would do well to get in the general vicinity of the correct building. This implied that with this new device they could blow open the front door.

Jones was wrong, but not so wrong that the report was raising unjustified alarm. The device's name? X-Gerät.

CHAPTER 7

The Spectre of the Brocken

Koventrieren, verb: to reduce to rubble

October 1940

THE BROCKEN IS THE tallest peak in central Germany. It isn't very high – not even as high as Ben Nevis in Scotland. But on an overcast day, as the mists swirl around the damp, forested valleys below, it can feel very remote, and walkers can feel very alone. This is when, just occasionally, there is a chance they will look down into the clouds below and see the haloed image of a grey man who seems to be following them.

The Brocken Spectre was first recorded in the eighteenth century. Johann Silberschlagg, a naturalist, was out hiking and saw, amid the broken mist and fog, a shadowy man ringed by a ghostly rainbow in the clouds below. He, though, was merely the first to write it down. Throughout the valleys, there were locals who had claimed to have witnessed the same ethereal, ominous presence. Actually, many had throughout the world. But there is something about the weather patterns on the Brocken that makes this spectre far more likely to emerge. The Harz mountains, of which the Brocken is the highest, form a barrier across central Germany. Here the warm wet air of

Central Europe is pushed up into the cold – fog and rain are never far away. Neither, though, is the sun, burning off the morning mist in a daily battle with the clouds.

If you are standing on a ridge and the conditions are just right, there will come a point when the mist rising silently through the dank forests meets the sun blazing from above. That is when the spectre appears: not a ghost at all but, we now know, a rare optical illusion. If you are lucky, and standing in just the right place, you will see your shadow projected on to the clouds below – a spectral presence ringed by a full circular rainbow.

The Brocken Spectre has frightened hikers, intrigued physicists and attracted tourists. In the autumn of 1940, when the Blitz was well underway, it was to exert its influence again – in Reginald Jones's recruitment plans.

By now bombers were arriving nightly in great armadas, setting cities ablaze. There would be rumours, though, that something worse was to come. There were rumours of an operation so devastating, against such a key industrial city, that it would test the mettle of the nation itself – that it would test the theory, long posited by both sides, that bombing alone could break a country. Details were scant, but as the autumn progressed Jones gained confidence that, somehow, this mysterious device known as X-Gerät was involved, the device he had once thought was merely a synonym for Knickebein, and he intended to find out how. When he was told that he could hire an assistant to help him, his mind turned to the Brocken Spectre.

In normal times, Jones might never have been offered an assistant. He might not have even had a job. There was good reason for this. The problem with Jones was that he annoyed people. Primarily, this was because he was just really annoying; he was brash, arrogant and self-regarding. He boasted, he played practical jokes, he had a high opinion of his own abilities.

On his scientific reports there are marginalia to be seen, written by his superiors, including furious notes that express growing exasperation with this young intelligence officer. In one of the surviving copies of one report about the X-Gerät, you can see the reader becoming increasingly incensed by the certainty of Jones's pro-nouncements, as well as his constant (and correct) implications that British aerial navigation methods were not themselves up to much.

As the report goes on, these notes – probably from Group Captain Lywood, Deputy Director of Signals – become first terse, and then querulous, and then finally outright furious. By the time Jones's report embarks upon an imaginary dramatized conversation between two leading German radar scientists and several officers (in which all Nazis concerned seem to be remarkably in agreement with Jones's own views), it is clear that this particular reader has finally lost his patience. 'This sort of play acting in a report of this nature, if advis-able at all, would be more useful if the characters had some real substance of thought!' he writes, testily. That short-tempered annota-tor was not the only one to be irritated by Jones. After the war, Solly Zuckerman, a scientific advisor on bombing strategy and, later, the first chief scientific advisor to the British government, described Jones as a 'silly man' – all but calling him a self-publicizing egotist.

But his superiors kept him on. In peacetime, being amiable but wrong can be a reasonable career strategy. It isn't in war, as few knew better than Jones himself. 'Despite any unpopularity, I survived because war is different from peace: in the latter fallacies can be covered up more or less indefinitely and criticism suppressed, but with the swift action of war the truth comes fairly quickly to light,' Jones later wrote, with characteristic lack of modesty.

If there is one thing more infuriating than arrogance, though, it is justified arrogance. If Jones was self-regarding, he had good reason to be. So much so that, by October, there appeared to be a grudging acceptance among British high command that he might even be use-ful enough to deserve an assistant. Or, as Jones put it, each night he was advising Fighter Command on where to put their planes to best counter the Germans flying in to attack Britain. And, he also pointed out, he was the only man in Britain who was able to do so. Therefore, if he was 'eliminated by a bomb, there was absolutely no one else who knew the technique'. Even the British could see that that would indeed be most inconvenient.

The truth is, though, Jones also just needed help. Knickebein was under control. He knew what it was, and how to counter it. But increasingly worrying reports were coming through about this other system, also clearly radio-based, called X-Gerät. Some of the most troubling originated with Denys Felkin and his prisoners at the Cockfosters Cage. Men, especially young men, like hot new gadgets. No gadget was hotter or newer in the autumn of 1940 than the X-Gerät – the system that British intelligence had once thought was merely another name for Knickebein. The prisoners soon made it clear it was something quite different. X-Gerät was equipment, explained prisoner A704, 'which is just like Knickebein but much more accurate and much more sensitive'. 'The beam,' he said admiringly, 'has an edge like a knife, so you can fly along it perfectly with this marvellous apparatus.' Although, they knew to keep such details from their captors. Such was the secrecy, another explained, 'You get your head chopped off for even talking about it.'

The rumours, nevertheless, had spread throughout the Luftwaffe. Instead of a fallible human ear listening out for the beam there was, one prisoner claimed, a clockwork system: a clock hand wound down, and when it hit zero the bombs released. Another talked of a 'pointer' that 'oscillates' to tell you when you are on the beam. 'You must keep flying to the left if the pointer points to the left. If you fly across the beam the pointer swings to the right.'

Even this was not the very best bit. Bored in his cell, prisoner A646 explained what he had heard second hand about the system to A629, and to the listening intelligence officers. The best bit, he said, is what happens when you are over the target. 'Suddenly . . . you find the beams. Crash! And the bombs drop.' 'Automatically?' 'Yes, by electricity'. A629 was suitably impressed. 'Automatically! Oooh!' There was no sign that anyone twigged they were being listened to. It should be noted that A629 may not have had the most informed of teachers. The rest of the conversation was sufficiently disconnected from reality that the interrogators listening in felt moved to add a note to their report stating, 'A646 knows really very little on the subject.' Already, it seems, the men and women employed to listen to the Germans were far better informed on German gadgets than most of their subjects.

Not, however, informed enough. To stop this new X-Gerät, whatever it might be, Jones knew he had to understand everything about it. He had to know it better than the pilots, better than the radio operators, better than the engineers. He had to know it well enough to spot weaknesses they had not even thought about. And he had to do so fast. Amid the breathless descriptions of this device, the Luftwaffe's new toy, was the realization that it would soon take over as the main precision-bombing tool: designed to at last bring Britain to its knees. Who could Jones get to help him piece it all together, now that he was finally allowed a colleague? That was when he thought of someone almost as annoying as him, a man who ten years earlier had helped him to gas Oxford.

When it came to research, he and Charles Frank, a chemist, had gone their separate ways. But their paths had still consistently crossed. Both, initially, had stayed at Oxford. Frank, who learnt Jiu Jitsu, took to dropping in at Jones's digs to practise his latest hold. At other times, he would turn his attention towards any gadgets or equipment in Jones's room – taking them apart methodically and then seeing if he could put them together, not always successfully. Jones took to hiding his equipment. The two found they were philosophically matched, as well as scientifically. Both had little time for people they viewed as talking nonsense. Occasionally, Jones said, Frank would 'surprise us by attending a religious service . . . which he did – he explained to us – to reassure himself that he was right in not accepting religious dogma.'

There was something else on Frank's CV besides science and martial arts, that would prove extremely useful in the years to come. In 1936, he had headed off to complete his PhD in the Kaiser Wilhelm Insitut für Physik in Berlin. It was here where he had also experienced his first period of freelance employment for British intelligence. In the middle years of the 1930s, the Air Ministry in the UK had become concerned about reports of a large aerial that had been erected on the Brocken, ostensibly by the German Post Office. According to the more lurid accounts, it was a prototype super-weapon, of the kind British scientists had argued was unachievable. So powerful was it that passing cars found their engines no longer

worked while it was switched on. Had the Germans perfected the
death ray? Did they have a device that could cause planes to drop out
of the sky – rendering their electronics useless?

In 1938, at the height of the Munich crisis, this was all too worrying. Jones, who by now was working in a temporary capacity for Air
Ministry intelligence, as he awaited transferral to the Admiralty,
thought that Frank, who was still in Berlin, would be the perfect person to investigate. But how? Just as MI5 was intercepting the letters
of Thost, so too was there a risk that Frank was having his steamed
open. Jones could not simply instruct him to go and look at the
Brocken – or the game might be up. They had to find a coded way of
telling him to go there, and investigate the area.

Then Jones remembered the Brocken Spectre. This was the cover
they needed. Jones was a physicist, this was a physical phenomenon:
what would be more natural than his asking about it? 'I wrote to him
suggesting that he might be able to give me a better description than
that available in textbooks,' said Jones. 'Immediately realizing that I
must be after something, he promptly burnt the letter and paid a visit
to the Brocken.'

There, Frank inspected the tower and wrote a description of it. He
also sent back a photograph to Jones. His car did not stop, he confirmed to no one's surprise but everyone's relief. In fact, Jones later
wrote in an official report, there was probably a simple explanation
for this rumour. As Frank described the tower it appeared that part
of it was a highly sensitive receiver: sensitive enough that any electronic noise, such as that from a car ignition, could interfere with it.
'This problem was simply solved by the Totalitarian method of placing guards on all roads to stop cars from proceeding' during
experiments, wrote Jones. 'It required very little distortion of the
actual story for it to be related that the cars stopped before the soldiers appeared out of the wood, thus giving rise to the rumour of an
engine-stopping ray.'

Jones forwarded all the details that Frank had sent him to the Air
Ministry and received, he said, a gratified response. 'It transpired
that between us Charles and I had produced a better and prompter
description of what was actually happening on the Brocken than the

MI6 agent who had been briefed to follow up on the original report,' Jones wrote.

But poaching such a useful scientist in the autumn of 1940, especially one who, following the posting, could speak German, was not easy. By this time Frank was, like every half-decent scientist, employed in vital war work – in his case at the Chemical Warfare Experimental Establishment in Porton, today known as Porton Down. Frank's employers were, it turned out, not keen to give him up. While Jones pulled strings and wrote pleading letters, the clock was ticking. A big raid was imminent. At the same time, though, the details slowly fell into place of the system known as X-Gerät. A name he had once thought was a synonym for Knickebein was, he now realized, something entirely distinct.

Sitting in his office, behind the incongruously imposing facade of the Minimax Fire Extinguisher Company, Jones had no single, perfect source from which to make deductions. He could only interpret, and guess. Like a lost Dornier, alone in the dark, he could triangulate from fixed, known sources: from the Y-service staff, those British radio enthusiasts who, whether in sheds on top of poles or vans along the south coast, listened out for German radio transmissions; from photographic reconnaissance flights conducted by the RAF, who had pinpointed the source of those signals; from German POW conversations, surreptitiously recorded through the walls; from Enigma intercepts of German messages. Also like a lost Dornier, he had to hope that these signals were true – that they did not come from unreliable sources or, worse, purposely deceptive ones. He was the sole, lonely navigator of Britain's scientific intelligence.

The job of an analyst in an intelligence agency is to draw together all the sources, weigh them, interpret them, and synthesize them. But in a profession that preferred classicists to chemists, he sometimes felt very alone. As he tried to imagine the thought processes of his opposite numbers, all he had was educated inferences. He did not know what those young crews felt as they saw the pointer of their receiver indicate they had hit the beam. As they found the beam and headed into hostile territory, as they felt fear, or perhaps pride, he had to put himself into the mind of the crews and the engineers who designed the system they used. In particular, Jones had to work out why X-Gerät had not one intersecting beam, like the Knickebein, but three.

Slowly, tentatively, like the first faint dots and dashes on the ether, a route through the mystery started to become clear to Jones. In many ways, the prisoners whose reports he had read had the gist of it correct – even prisoner A646, the hapless inmate derided for not knowing very much. X-Gerät was indeed like a better version of Knickebein, although that description doesn't do justice to its sophistication.

HOW X-GERÄT WORKS

Step 1

On a typical sortie in September 1940, a plane with the X-Gerät system installed would take off from Nazi-occupied France, and navigate into the Channel. There, it would hear the familiar dots or dashes, the hidden, comforting signal – sent much less than a thousandth of a second ago, from a place a thousand times safer. This, though, was only the start – this was the 'coarse' beam, used to guide it into roughly the right area.

Step 2

Once the bomber was aligned, the true X-Gerät beam would take over. Its far shorter wavelengths created a far tighter beam. It was tight enough, Jones calculated, that to track its path it was important to take into account not merely the curvature of the Earth, but also that the Earth is not perfectly spherical – it is squished.

Without correcting for that, you got entirely the wrong path.

Step 3

For such a precise beam the Luftwaffe did not rely on a listener to interpret the dots and dashes. Instead, they created a visual display, the pointer the POWs had been talking about. If the needle pointed to the left, you needed to steer to the right. If the needle then

The X-Gerät's instruments allowed pilots to follow a tighter beam

switched to point to the right, you had gone too far and needed to steer left.

This display had the added benefit of making it far harder to jam. You can fool an ear with electromagnetic fuzz. You can confuse it with a similar note. You can mask the sound it is looking for. It is not so easy to fool electronics, to fool the precisely calibrated circuitry carried on these planes.

Step 4

The bombers are now on an electromagnetic highway into industrial Britain. But they still don't know where to release their bombs. For that, they need the cross beams – three of them.

The first beam is just a warning. It tells them to get ready. Fifty kilometres later comes the second beam.

Step 5

When they hit the second beam, a stopwatch starts. As it counts out the seconds, it is crucial the plane flies at a constant speed: the stopwatch was to be a speedometer.

In the air, Second World War speedometers were usually inexact. At best, they gave you a rough approximation of your air speed – your

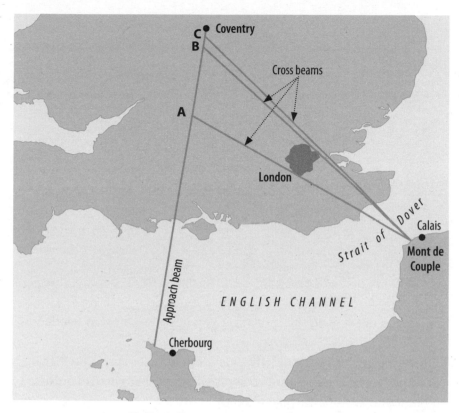

X-Gerät beams set to target Coventry

speed, plus or minus wind speed. What this speedometer gave was something far more valuable: land speed.

When, exactly 15 kilometres later, the plane finally passes the third beam, the stopwatch stops, and the reading is the exact time taken for it to cover 15km on the ground.

Step 6

Now, the stopwatch starts running in reverse – counting down instead of up. With a pre-programmed gearing mechanism, the engineers are able to alter the speed of this countdown before flight.

If the bomb release point was exactly 5km away, it would run three times faster – so that the 15 kilometres it took to wind up is now wound down in five.

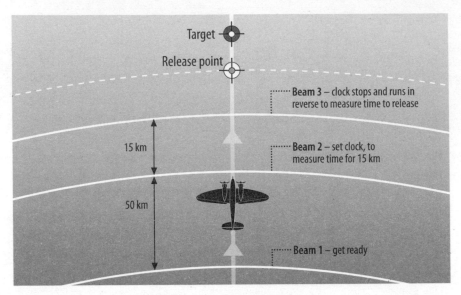

X-Gerät used three cross beams

But the genius of the system was it could be tweaked further, to account for the bomb ballistics and altitude of that particular plane, all of which modify the release point.

Then, when the stopwatch reached zero, the plane was, in theory, exactly where it needed to be, assuming it had not altered its speed or heading. An electrical circuit was completed and the bombs were dropped.

None of this was laid out in detail for Jones by one single source. It was a jigsaw, with pieces mislaid, misshapen and missing. Some pieces slotted in easily, a new nugget of intelligence at last providing clarity in a particular section. Some came from Jones's mind alone – from time spent sitting and thinking, trying to imagine what he would do if he were a German scientist faced with the problem of directing bombers over Britain.

Writing to his superiors, it was clear to him even before he had established all the details that they had perhaps underestimated the seriousness of their adversary. By defeating Knickebein they had simply lopped off one of the Hydra's heads; now another, more formidable beam had regrown in its place. He then went on to warn that while his information was still partial, that was no excuse to not take notice. 'This report may have appeared scarcely credible,' he said. But his warnings, he said, had form, even if they had been ignored. 'On some previous occasions, [they] have borne a similar history to those of Cassandra. It is hoped that in the present instance they will be taken seriously.'

It was indeed hoped. Such seriousness was not, though, uniformly in evidence. Developing electronic countermeasures was hard, took time, and required details about precise frequencies which Jones still did not have. Why bother, though? The simplest way to deal with the beams would have been to bomb the transmitter sites. They knew the exact locations – why not just obliterate them?

One answer is that doing so involved explaining the subtleties of this quite complex, scientific system to Arthur Harris. And Harris, better known as 'Bomber' Harris, was not a man for scientific subtleties. On being asked to divert some bombers from the task of bombing German cities to the task of destroying these aerials, his response was blistering – and for those involved in combating the beams, mortifying. 'Are we not tending to lose our sense of proportion over these German beams?' Harris began. 'We use no beams ourselves but we bomb just as successfully as the Germans bomb, deep into Germany,' he added – several months before the RAF's operational research would show just how unsuccessful our bombing actually was. 'Lack of beams will not stop the Boche,' he said, 'and in my opinion will not even embarrass him.'

Harris had another objection: Even if bombing the aerials in occupied Europe was sensible, the frightening attrition rate being suffered by British bomber crews at this point in the war made it impractical to train them up for the task. 'We are always liable to forget that the average life of a bomber crew is three to four months,' Harris said. 'Special training . . . is inevitably rendered nugatory after a few weeks by changes in personnel.' But, Harris added, warming to his theme, this was by the by. 'I do not agree that the beams are in fact a serious menace to this country or that they have proved to be in the past.' He would, he added, 'go further and say they're not even really useful.'

'I quite agree that "bending" them has given us great amusement and the Boche a good deal of annoyance and possibly loss.' That, though, was all. In Harris's view, the beams were a waste of time for the Germans, and a help to the British because they gave Britain's intelligence officers forewarning of what targets the Germans were aiming at.

Harris concluded with the sort of flourish of which General Melchett in *Blackadder* would have been proud. 'We are not such fools as the Boche! . . . Long may the Boche beam upon us.'

CHAPTER 8

On Chesil Beach

'This is how the entire course of a life can be changed:
by doing nothing'

IAN MCEWAN, *ON CHESIL BEACH*

*7 October 1940. Five weeks before the rumoured raid on a
Midlands city*

FELDWEBEL OSTERMEIER WAS HAVING a bad month. This was
not, alas, an unusual feeling for him. The sergeant-major was in
charge of one of the X-Gerät beams, and it was finickity work. To
project a thin beam of synchronized electromagnetic waves for hundreds of kilometres a lot had to go right. The corollary was, a lot
could also go wrong. And, for Ostermeier, more often than not it did.

His was not one of the offensive beams, tracing a route over Britain. Instead, it was used for practice – projecting over central France.
Even so, this was a time when a lot of practice was needed. His beam
was crucial for allowing bomber crews to learn to use X-Gerät in
safety. And, if he was entirely honest, he wasn't very good at his job.
Or, as Jones put it, in what was ostensibly a serious technical report
on his ongoing X-Gerät investigations, '[Ostermeier's] misfortunes
and plaintive garrulity rival those of Donald Duck.' Despite the

report being circulated to many of the most powerful men in the country, most of them twice his age and ten times his seniority, Jones couldn't help but fill half a page painting a vignette of the hapless Ostermeier. 'He seems to be well known throughout the whole company for his comic stupidity and those in charge of the other stations delight in sending him dud equipment,' he said.

October was, said Jones, particularly unfortunate for Ostermeier, as 'he started it by receiving a complete transmitting set which was defective in so many ways that no one piece of the apparatus could be tested individually. He then waged an epic struggle in which he no sooner put one component right than another went wrong.' Worse, whenever he ordered components from the military stores, they too failed to work. For one glorious day he got the whole set working before – Jones can barely hide his glee – it broke down again. Throughout, poor Ostermeier described 'his plight in somewhat euphonious letters to the stores'.

By this time, the London Blitz was at its height. Jones was exhausted. He had had his sleep disrupted seventy nights in a row. Amid the destruction and the high stakes of his job, Ostermeier was a rare bit of comic relief. He was not to last. Goering was taking increasing interest in the X-Gerät, his latest superweapon. 'And,' wrote Jones, 'the presence of the fabled Ostermeier was not thought to be conducive to maximum efficiency.' Ostermeier's superiors took action. 'It was with sincere regret that we heard of his passing to a subordinate position.'

The fall of Ostermeier was a sign that Knickebein, too, had passed to a subordinate position. There was no place for people like Ostermeier in the X-Gerät programme any more. X-Gerät was set to take its place at the tip of the Luftwaffe's spear.

14 October 1940. Four weeks before the raid

In Reading, Berkshire, the local government Regional Information Office had prepared a strategy for dealing with the long-expected mass bombing of civilians. As the summer of 1940 had come to an end, a summer many spent lying in the fields of Kent watching the dogfights above, the Luftwaffe had changed tactics. The Blitz had begun in earnest, civilians would be the target and, in the information office, they thought they were ready.

In order to keep up morale there were, they explained in a brisk memo, six arguments.

1. The patriotic: 'We must be British and stick it; men in the Forces have to put up with worse than this.'

2. The statistical: Comparison with road casualties, the area of England and so on.

3. The retaliatory: 'Worse things are happening to Germany.'

4. The fatalistic: 'If a bomb's got my name on, it will get me.'

5. The shame argument: 'We should be ashamed of ourselves for behaving like this.'

6. The stubbornness attitude: 'If the Germans think this is going to [get] us down, they will be disappointed.'

In Britain we still invoke the 'Blitz spirit' today. It is a term for pulling together in adversity, for facing down hardship with community spirit, doggedness and not a little stiff upper lip. As September became October, as night after night the bombers still came back, there was a lot of Blitz spirit. The national upper lip certainly had its elements of admirable stiffness. But there was some quite understandable wobbling too.

The East End of London was the first to be hit. With the Thames tracing a silvery and meandering line through the capital, it was a target that required little navigational skill. In Deptford, on 10 September, a Home Intelligence report was alarmed by the sight of refugees, of families 'making for the hopfields of Kent, taking with them such of their belongings as they can carry . . . without any other apparent object than "to get away from it all"'. In the 'Dockside area', they warned a little later that there were 'visible signs of nerve

cracking from constant ordeals'. 'Old women and mothers are undermining the morale of young women and men by their extreme
nervousness and lack of resilience. Men state they cannot sleep
because they must keep up the morale of their families.'

Again and again, the issue of sleep comes up. The casualties were
horrific – more than five thousand dead in the first month in the
London area alone. But there was also the exhaustion. Entire cities
were unable to get any rest. They were, Home Intelligence feared,
close to breaking. 'When siren goes people run madly for shelter
with white faces . . . Conditions of living now almost impossible and
great feeling in Dockside areas of living on island surrounded by fire
and destruction.' They began looking for enemies. Rumours began,
again, of fifth columnists. Some said the searchlights were guiding
the bombers in. There were antisemitic attacks.

We know that morale will hold even in the most desperate of circumstances. Even so, across the Channel there was an invasion force.
In the air above, there was a bomber force. Britain was fighting for its
existence, and everyone knew it. This was not war as had been known
before: fought in foreign fields, far from home. This was total war,
that intruded – or threatened to intrude – on every inch of England's
green and pleasant land. On the south coast, Britons peered into the
sea mist to watch for the dark shapes of landing vessels. Throughout
Dorset and Kent they saw the defensive lines, prepared inland – on
the assumption they would be occupied. Further inland, people
watched the skies for the hairy-handed nuns who would parachute
in on a moonlit night, or for the bombers that could rain down death
at will. And while the nuns may not have come, the bombers did. As
one East End resident put it, 'It was the most amazing, impressive,
riveting sight. Directly above me were literally hundreds of planes,
Germans! The sky was full of them. Bombers hemmed in with fighters, like bees around their queen, like destroyers round the battleship.
So came Jerry.'

Technology had brought war to the skies above Britain; hidden
among those swarms of planes was another technology, guiding the
planes to their targets. In this war, a new kind of technological battle,
good scientists were at a premium and if you had one, you clung on

to him or her. So Charles Frank's superiors at Porton Down had not taken kindly to letters asking for him to be released 'for some extremely important and urgent work'. They very reasonably considered the work Frank was already doing to be 'extremely important and urgent'. He was living in Wiltshire and working on research into smoke generators that were designed to produce billowing clouds of smoke to hide factories from the air.

He was also, like Jones, married – and his wife, Maita, was not keen on moving to bomb-riddled London. 'Please don't think her too cowardly', Frank pleaded, in a letter to Jones. 'I think it probable that your job would be more important than what I do here, and more interesting ... you have put me in rather a quandary ... I should come without hesitation if unmarried, but now I am uncertain.' There was an additional complication, he added. Even if he persuaded Maita, did MI6 want a foreign national's husband in its office?

When Russia had invaded Finland, in 1939, Frank had taken up the cause of the Finns, fundraising enthusiastically in his new home of Cambridge. Someone on the committee had suggested it might be a good idea to find an actual Finn to give a talk, and Cambridge had been scoured. Two were found, both women. One was middle-aged, the other was young and attractive. It was no contest which they would invite. Maita gave her talk, and Frank was smitten.

She was 'a friendly alien, but an alien all the same', explained Frank. This could cause complications. After some days of deliberation, the pair decided they would risk London, and Jones was more than happy to risk Maita. Charles packed up and came as an advance guard. Until he could find somewhere of their own, he moved into the Joneses' cramped Richmond flat.

Three weeks before the raid

By the time Maita and Charles had settled down in the Joneses' spare bedroom, the amount of detail Jones had gathered about how the Germans used X-Gerät to direct their bombs, and about X-Gerät itself, was stunning.

He knew the Germans typically chose their targets in the afternoon of the attack, and he knew that the precise coordinates and
settings of the beams were calculated shortly after. He knew that for
some beams this information was then communicated from headquarters to the beam operators themselves by wireless, whereas for
others it was sent by landline. He knew that the masts transmitting
the beams across England from occupied Europe were mounted on
turntables, so they could be turned and directed to new targets as
required. He even knew that for the newer machines that the Germans had installed, this turntable was octagonal and could be rotated
through 360 degrees. He knew the two different kinds of silenced
petrol generators used to power the transmitters and he knew that
the machine's regulator came from the company Julius Pintsch. He
knew the valves that were installed, the staffing ratios of officers to
subordinates at the beam stations, and that a van drove to around a
kilometre away in the direction the beam was pointing to listen out
for it and make sure its trajectory was precisely aligned. He knew
that Ostermeier, before his fall, had wanted to get heated huts built
for the staff of his monitoring vans, so they didn't get too chilly waiting as winter arrived.

There was one other bit of information he gathered, arguably even
more crucial than the fate of poor Ostermeier. It was that the short-
age of receiving equipment meant that just one of the elite flying
units that was currently attacking Britain would be using X-Gerät.
Unlike Knickebein, in which all planes had receivers and could hear
the beams themselves, most of the German bombers flying in each
night would have no receivers – they would have to be led to their
target by another unit.

This one German unit – three small squadrons of highly specialist
planes – would be employed to 'paint' the target within Britain – to
set that particular town, or factory, or city on fire, with incendiary
bombs, so that it would blaze as a beacon over the horizon. By creating fires the pathfinders would light up the sky and draw in the
following stream of bombers. This was the pathfinder unit called
Kampfgruppe 100. The use of pathfinders marked a new kind of

aerial warfare, that would be taken up enthusiastically by both sides. Preventing them finding their target still represented Britain's best course of defence, even if most of its citizens didn't realize it.

Home Intelligence had reported a big boost to morale when more anti-aircraft guns were brought in to London in mid-September. The noise, they wrote, 'brought a shock of positive pleasure'. In the shelters, people cheered. 'It made people feel that all the time we had a wonderful trick up our sleeves ready to play when the time came.' Trick was the right word. Jones was under no illusions about the purpose of the barrage: it had more to do with the morale of the people on the ground than with the bombers above. The fire was too inaccurate; the bombs would land irrespective. The bomber, as Baldwin said, always gets through. Britain still could not effectively fight the planes, from land or in the air. Instead, the only way that Britain had of protecting her cities from the nightly fleet of attackers guided by X-Gerät was to confuse them, to divert them from their targets.

Doing this required more than merely knowing in general terms how the X-Gerät beams worked. Jones had enough information to have given a lecture to the air staff describing, in theory, the entire apparatus. What they needed, though, in order to disrupt the beams and confuse the planes, were the precise specifics. What frequency were these beams being transmitted on and how? Every day the German command communicated the settings for the equipment in the planes, telling the planes the frequencies they needed to listen in to – tuning in just like you would to a particular station on a car radio. If the radio operators were on the wrong frequency, they wouldn't hear it – just as someone who wants Radio 3 but is tuned to Radio 4 gets the afternoon play or *Thought for the Day* rather than Wagner. And if the British blocked the wrong frequency, then they too would end up disrupting a message the Germans were not listening to in the first place.

If Britain was to survive the Blitz, and come out of it in a position to retaliate, then Jones and Frank needed to make sure that when the bombs landed it was somewhere useless – that they churned up the soil of this green and pleasant land more often than its mills. Increasingly, it looked like doing so meant neutralizing the elite pilots in K.Gr. 100, and their secret weapon, the X-Gerät.

Two weeks before the raid

To understand how important K.Gr. 100 were it is only necessary to run through the Enigma intercepts describing the different dignitaries coming to visit them. It was clear to Jones that they were regarded as 'the most distinguished Gruppe in the German Bombing Force'. Even so, it was very hard to tease apart the effects of the group in the UK. Jones needed to work out just how good they were. Throughout the summer and early autumn, they were working on their own – a small, precise force. How to spot the patterns of their precisely targeted bombs, amid the thousands of other less accurate raids happening each night?

The trick was to follow the beams. Thanks to intercepted messages, instructing the beam operators, it was possible to work out where they had crossed in previous nights. By correlating these crossing points with the reports of raids, it was possible, Jones found, to retrospectively chart their routes, and their efficacy. This was how Jones realized that while he had been concentrating on Knickebein, this one force had been quietly practising pinpoint raids. From where the bombs fell he was able, at last, to compare theory with reality. 'It was stated,' he wrote, referring to his own preliminary report, 'that the accuracy expected by the Germans appeared to be something like 10–20 yards over London.' This was based purely on the accuracy with which the Germans set up their beams – though of course human error in flying, not to mention changes in wind speed after bomb release, would chip away at this.

The estimate, continued Jones, 'has been at least partly justified by the performance of K.Gr. 100 over Birmingham towards the end of October . . . Their most spectacular effort occurred on 26/27.' On this occasion, three sticks of bombs were dropped, on average falling a bit over 100 yards from the line of the beam. Later in the war, it became clear that if you sent a hundred British bombers using conventional navigation to bomb a factory, they would have been genuinely lucky to get within a couple of miles of their targets. Here was a device that would in theory ensure that with just a few bombers and a factory-sized target you would be genuinely unlucky to miss.

If it was just K.Gr. 100 they had to worry about, this would be annoying, but not war-changing. Another tactical shift was under-way. K.Gr. 100's pilots were not glad-handing generals because these couple of dozen planes were the most technologically advanced in the bomber force, although they were. No, the point was that they were preparing to be the leaders – to blaze a trail for all the others.

At the start of November, there was a large bump in German signals traffic, frantic chatter between airfields, beam stations and headquarters. Britain had already been under two months of almost continuous bombardment. Now it was clear that something bigger was coming, a major raid different in scope and – Jones suspected – accuracy. It was to be led by K.Gr. 100. For Jones and Frank, the race was on to be ready. They had to crack X-Gerät.

With the nation facing the throbbing migraine of X-Gerät, some-thing altogether more powerful than Aspirin was required. Logically enough, this new system was called 'Bromide'. There was no time to wait for all the details before beginning the development of Bromide. Robert Cockburn's team at the Telecommunications Research Estab-lishment had begun work on the jammers immediately, adapting portable army radar to the task – using whatever information Jones could give them. As Jones's understanding of the system was refined, so too was the jammers' design, the countermeasures tweaked on the hoof.

It was not enough merely to be close to the answer. Such was the sophistication of X-Gerät that jamming it required absolute preci-sion. Over the course of two hectic months Jones and, latterly, his 'staff' – he and Frank – had gathered the plausible range of frequen-cies they needed to cover, the strength of the signal they had to jam, and the orientation of the different beams involved. Sometimes this was just about collating the information. Often it involved genuine detective work and deduction. Take, for instance, the 'Anna' number.

Through Enigma intercepts they had worked out that the 'Anna' was some kind of radio device carried in the planes that were using X-Gerät. On this 'Anna' device they knew that there was a dial, and that dial was labelled with numbers denoted in 'grad'. But what did the 'Anna grad' mean? How did these apparently arbitrary numbers relate to the real world? From a chance remark by a German officer came the information that the numbers represented a frequency – that this was, presumably, how the planes locked on to the right beam. That was one clue.

What frequency, though? These were arbitrary numbers. The volume dial on a stereo might go from 1 to 10, but the numbers are just a convenience – each corresponding to the true decibel output. The same was true with the Anna numbers. Each related to a frequency, but how they related was a decision made by the manufacturer. If you knew that K.Gr. 100 were planning to lock on to a beam with an Anna grad of 25, what actual frequency was that?

The way in which Jones calculated this key information is a good example of his role as detective, as a Sherlock Holmes of radio. It is also ever so slightly complicated, so if a mathematical diversion is unappealing then skip this section.

From intercepted messages, Jones knew some of the different Anna numbers used in real life – sent to beam stations and bomber squadrons to synchronize their settings. They were 10, 15, 25, 30, 35 . . .

Here was another clue: aside from a couple of exceptions that could plausibly be errors, each number, Jones noted, was a multiple of five.

From the Y-service, who listened out for enemy transmissions, came the real world frequencies used on the beams. 66.5 megacycles per second, 67, 67.5, 68, 69 . . . Each, Jones also noted, was between 65 and 75 and was a whole or half integer. That was also a clue.

So Jones knew the Anna numbers went up in increments of five, while the frequencies they mapped to went up in increments of 0.5.

His hypothesis was: if the first set of Anna numbers – 10, 15, 20 – related to the second set of frequencies – 67, 67.5, 68 – then there should be a simple formula linking them. The first step to deriving it was obvious. Frequencies went up in half integers and Anna numbers went up in fives. Clearly, the first step was you should divide the Anna number by 10. That was not enough, though.

If you took the Anna number of 10 and divided it by 10 you got one – a number very far from the 65–75 million cycles per second they were picking up.

So it also seemed sensible to presume that after dividing your Anna number by 10 you should add it to or subtract it from a constant: a larger, fixed number.

Or, as Jones put it: Frequency = Constant + (Anna Number)/10

For instance, if they knew an Anna number of 20 meant a frequency of 68, then that must mean the constant was 66, because 20 divided by 10 is 2, and 2 added to 66 is 68. If you know that, then you also know what all the other Anna numbers mean. An Anna number of 35 would be 3.5 added to 66, which was 69.5.

This was key. Get that constant, and you could relate the night's Anna number to the precise beam frequency.

But what was the constant? Here came the most cunning deduction of all.

One night, intercepted communications showed that a beam station worker called Schumann signed a receipt for radio equipment corresponding to 69.5 mc/s and 70 mc/s. Then, in mid-October, he was told to transmit on Anna 30 and Anna 35. That only worked if the constant was 66.5.

So it was that, thanks to prisoner interrogation, radio interception, Enigma decrypts, and some clever deductions, it was possible to know what frequencies were being used just by knowing what settings the bomber's Anna was told to tune to. This was just one deduction among many, one small piece of the puzzle.

6 November 1940. *Eight days before the raid*

From above, Chesil Beach looks so invitingly similar to a runway. An 18-mile spit, 200 yards wide, it runs straight and true from Bridport Harbour to Portland. On a moonlit night, the shingle gleams white, a sharp barrier between the dark fields of Dorset to the north, from which it is separated by a lagoon, and the dark waves of the English Channel to the south. For a desperately lost bomber, seeking out somewhere to land in late autumn 1940, it must have seemed the obvious place to ditch. But as inviting as Chesil Beach looks from above, it is less so when it is coming up to meet you at 100mph. Rather than the hard sand the pilot would have hoped for, there is shingle into which the landing gear cut deep grooves. Rather than being the flat and level surface that it looks like from the air, it is in fact riven by ledges where the tide has pushed the stones into piles fifty feet high.

The Heinkel III bomber that crashed down on this beach on 6 November 1940 was a member of Kampfgruppe 100, the elite German pathfinding squadron. Which was rather embarrassing because, at this point, it had definitively lost its path. The navigator's state-of-the-art equipment was contradicting itself – his compass was showing different bearings to his new radio direction-finding equipment. Almost certainly, he concluded (too late), the radio was being interfered with by 80 Wing's meacons. So it was that, running low on fuel and still over Britain, his pilot had little choice other than to land.

In the crash, one member of the German crew was killed. The others, shaken, lacked the presence of mind to destroy the new device they had been given. Kampfgruppe 100's symbol was the Viking longboat – the vessel that a millennium earlier had navigated the featureless expanse of the North Sea to inflict terror upon, and eventually invade, Britain. The elite K.Gr. 100 was intended to do much the same as those ancient Norsemen. In a week's time, the unit was scheduled to lead a raid that would not only be the most ambitious of the war so far, but that would prove a new kind of warfare was possible. This raid, codenamed 'Moonlight Sonata', was a plan not merely to attack a city, but to obliterate it. From the ashes of the destruction would, they hoped, come proof of a long-held theory about aerial bombardment: that in the twentieth century the bomber, when massed in enough numbers, was on its own enough to crush a nation's will to fight.

By this stage, rumours of the raid had been circulating for some time, and Britain had a good idea where it would be – but did not know what to do about it. Within the wrecked plane on Chesil Beach was one of the devices, still intact, that would enable each plane in unit K.Gr. 100 to reach its target. Inside the wreckage of the plane currently lying on the shingle in the moonlight was the finest navigation equipment yet produced by humanity: the X-Gerät. This plane was, for British intelligence, a gift. In its electronics lay the crucial information that would mean the X-Gerät could be countered and the raid disrupted.

But there was a problem. Or, as Jones later put it, 'There then followed an episode as grimly humorous as the Porter's performance in Macbeth, and equally incongruous as a prelude to high tragedy.' In the early hours of the morning, when the Army arrived to cordon the plane off, they had found it lying on the very edge of the beach, the surf lapping at its wheels. They put a rope around it, to haul it in. But was this great prize, a rare example of a cutting-edge enemy bomber, really Army business? After all, isn't the sea, even the edge of the sea, the province of the Navy? The Navy certainly thought so. 'Up showed the Navy and said, "Oh, this is a naval task",' recalled Robert Cockburn, from the Telecommunications Research Establishment. It was, he added, 'one of these unfortunate wrangles' you sometimes get between the different branches of the armed forces. The Army said the plane was theirs. The Navy disagreed.

If all this wrangle had done had been to waste time, that would have been one thing. But as the Army and Navy argued on Chesil Beach about the limits of this liminal zone, the tide came in. The sea, notoriously unconcerned about interservice rivalries, splashed into the cockpit.*

Jones and Frank knew, now, the raid was planned for the following weekend, most likely the night of 14 November. They knew it was

---

* This story has attained something of a mythical quality, to the extent that the true sequence of events is disputed. In Jones's telling, the Navy arrived and, rather than contesting ownership, offered to help move the aircraft, already in the water, to shore. But the suspicious soldiers were under orders from their commander to let no one touch it, 'I don't care if even an Admiral comes along,' he said. 'You are not to allow him near it!' So they spurned the help. The end result is the same: the plane got very wet

going to be guided, with precision, by K.Gr. 100. They also knew it was going to hit one of three cities: Birmingham, Wolverhampton or Coventry.

Special beam instructions had been sent out, so everyone in the Luftwaffe who needed to be was ready in advance for the big day. So too was the RAF – the instructions had been intercepted. On 9 November, there was a name, too. An Enigma intercept outlined orders for an operation codenamed 'Moonlight Sonata'.

Would Britain's bromides be ready in time – would the deductions of Frank and Jones essential to their functioning be correct? There was one, final, crucial bit of information. And that bit of information was currently being slopped around in the cold and salty waters of the Dorset coast. Within that deteriorating plane lay the electronics that would tell them the modulation frequency, the 'note' that would be used to guide the German planes in to Britain.

A radio signal – whether a beam or a BBC broadcast – has a frequency on which it is transmitted. It also has, within that frequency, the information that needs to be sent – for example in the case of the BBC, the music or words being broadcast. And that information does not change depending on the frequency used. If an opera is broadcast on medium-wave radio and long-wave radio simultaneously, the singer doesn't sound lower on the latter broadcast just because her aria is transmitted at a longer wavelength. No, the radio picks up the signal and interprets it accordingly.

The same was true for X-Gerät. It had its frequency, and on that frequency it also had its 'modulation frequency' – the 'note' that was being transmitted, and which K.Gr. 100's apparatus listened out for. In order to properly jam the transmission, you needed both the frequency it was broadcast at and the precise note that was being sent along the beam. This is particularly the case if you are looking to fool not a human but a machine. A human might not notice a slight difference in pitch – but a machine would. How to determine this, the modulation frequency?

All this information and more now lay within the grasp of Britain's radio engineers. With the crash landing of the Heinkel, everything Jones and Frank needed was all there, a gift from the Luftwaffe. The fuselage of the bomber lay intact amid the shingle of Chesil Beach. The Army made their case, the Navy made theirs. Inch by infuriating inch, the waters rose. Time and tide wait for no man, and this was not the only place where time was against Britain.

The expected night for the big raid was 14 November. The plane had landed on the 6th. To stop X-Gerät required blocking the main director beam and at least one of the cross beams. The countermeasures were ready to do this, not a day too soon. With enough warning and a bit of luck, the scientists at the Telecommunications Research Establishment believed there were now – just – enough jammers that could be brought to bear. A week earlier, two had been positioned in the likely path of the beams. But speed was everything. The team at 80 Wing, who operated them, needed to set the frequencies correctly.

14 November – Operation Moonlight Sonata

Saturday morning was bright and clear in Coventry, and Jean Taylor was looking forward to her fourteenth birthday. There had already been several small raids on the city, and she was used to the routine. They had a spot in a shelter, and if the sirens were heard she would head down there with her parents and older sister. Her father had decided early on that she wouldn't be evacuated. 'We live together as a family,' he said, 'and we die together as a family.'

The afternoon of the 14th came, with the Heinkel on Chesil Beach still not properly examined. Neither had the team at Bletchley been able to provide intercepts of the Anna codes in time. The British knew that an attack was coming. They knew it was coming that night. They didn't know where. Unlike with Knickebein, the Germans only switched on the X-Gerät beams at the last minute, so the only option left was to send up a plane and listen out for the beams themselves.

On the night of the 14th, a plane was sent up from an airfield in the Midlands, with a set of listening apparatus. It flew parallel to the east coast, searching for – and at last finding – the distinctive signal of the X-Gerät. The results were telephoned to Jones, who was waiting in his office in Broadway. He took them down; and he realized that they were wrong. The frequencies had to be half or whole integers, but the necessarily fuzzy measurements of the listening service had numbers in between. One frequency, the caller told Jones, was 68.6 mc/s. Jones told him to set his jammer for 68.5. 70.9, Jones advised, was really 71. 'But deciding what, for example, 66.8 meant, was more of a lottery.' Should they jam at 67, or 66.5?

It was 6 p.m. on the 14th and there was no time for a second opinion. The fate of a Midlands city – the exact location was still being determined – rested on Jones's answer. He went for 67. 'It was a most diabolical bit of gambling, as you can imagine,' he said. 'Because if one's wrong perhaps five hundred people are dead in the morning.'

3 hours to go

In French airfields, the German crews prepared their planes. In coastal beam stations along the Normandy coast the German engineers checked their equipment. In central England, the new bromide jammers were calibrated and positioned.

On the roof of Coventry Cathedral, the Cathedral Provost, the Very Reverend Richard Howard, stood with three volunteers above the nave, with buckets of sand to put out any fires. It was a bright and clear moonlit evening, and beneath them the rooftops of historic Coventry shone bright and ominously clear. Beneath one of those rooftops, Jean Taylor's mum had packed a flask and sandwiches, in anticipation of another raid.

And in Richmond, Reginald Jones, Vera Jones and their house-hunting lodger Charles Frank went to bed. There was nothing left to do. Jones had, he said, 'an uneasy night', as he lay wondering what he would hear on the eight o'clock news the next morning.

The morning after

On what was once a street, amid still smoking rubble, came a dog – trotting through the debris. The destruction around it was near total.

In places, the city was still burning – the smell of charred flesh mingling with the ash.

The first bombs had hit just as the other people in Jean Taylor's air raid shelter had been singing her 'Happy Birthday'. Then, for eleven hours, they had sat together, unable to sleep. 'We knew this one would be different because it started early, about 7 p.m. It then went on and on and on,' Taylor recalled. 'We were packed into a shelter – a school – with three hundred people for eleven hours. It was horrendous. I was sitting there between my sister, who was really upset, and my brother-in-law who'd come back from Dunkirk shellshocked. 'I remember saying to God that I just wanted to make it to fourteen.'

On the roof of the cathedral, Rev. Howard and his small team had battled until they could battle no more – until they had no sand, no strength and no hope. Before midnight, the fight was lost. They left to save themselves; the cathedral burnt to the ground. The next morning, behind the charred remains of the altar, Howard wrote the words, 'Father Forgive'.

Something changed that night in Britain. It changed in the Luftwaffe too. In their nightly orders, the town had been described as 'an important centre of the enemy armament industry'. Looking down, that clear moonlit night, the crews did not believe it. 'The usual cheers that greeted a direct hit stuck in our throats,' one pilot wrote. 'The crew just gazed down at the sea of flames in silence. Was this really a military target?'

Throughout the city centre the next morning residents emerged from beneath the ground, from cellars and shelters under buildings that no longer existed. Shell-shocked and dazed in the dawn, they found bodies lying in the street. 41,500 homes were damaged, 568 people had died.

Area bombing was horrific, but it was not mindless. There was a theory behind it. Inflict enough devastation on a people, military strategists of the 1930s had hypothesized, and you could bring a nation itself down. In the population's desolation and demoralization lay victory.

Here in Coventry was an opportunity to test the idea, for both sides. Whitehall sent down observers to record the aftermath. Their

conclusions were grim. 'Women were seen to cry, to scream, to trem-
ble all over, to faint, to attack a fireman,' wrote Tom Harrisson, an
anthropologist. 'The size of the town meant nearly everyone knew
someone who was killed or missing. The dislocation is so total that
people easily feel that the town itself is killed. "Coventry is finished"
and "Coventry is dead" were the key phrases . . . The overwhelmingly
dominant feeling was utter helplessness. The tremendous impact of
the previous night had left people practically speechless in many
cases. There were more open signs of hysteria, terror, neurosis,
observed in one evening than during the whole of the past two
months in all areas.'

The dog continued its tour. In its mouth, witnesses later reported,
it carried the arm of a child.

Jones listened to the first reports of a raid 'on a Midlands town'
over breakfast. He was devastated. The frequencies, he assumed,
must have been wrong and he had been, he believed, 'instrumental
in contributing to the disaster'. Yet, later in the afternoon, 'my wretch-
edness turned to bewilderment'. Decrypted messages to the radio
stations showed his guesses had been right. The frequencies were
correct. What had happened?

He had to wait a week for the answer – or, at least, *an* answer. On
21 November, he visited Addison and the rest of 80 Wing in their
headquarters. There – wet, salt-damaged, but intact – they at last had
the full X-Gerät system of the Chesil Beach Heinkel. Robert Cockburn and the team at TRE had already examined it. They found there
was a 'very sharp . . . very accurate' filter in the equipment. That filter
was designed to listen out for the modulation signal of the X-Gerät,
and only the X-Gerät. It was set at 2,000 cycles per second – a high-
pitched note that would have sounded like top C on a piano. If the
bromides also played a top C, then their jamming would mesh precisely with the real signal, and the Heinkels' electronics would be
entirely unable to distinguish the real navigation instructions from
Britain's fake ones. But they were not playing a top C. 80 Wing's bromides had been playing a G. They had been playing 1,500 cycles per
second, and because of this as far as the radio receivers of K.Gr. 100
were concerned they might as well not have existed.

It should not have taken seized apparatus to spot this. The correct settings had been there for all to hear, a note projected over Britain more purely and clearly than by a King's College choral scholar in an evensong solo. But it had been missed, and the jammers had been useless. 'Whoever had made such an error,' said Jones, 'ought to have been shot.'

This tale will remain one of the great 'what ifs' of British history. As it happened, the 14th was a bright and clear night, with a strong moon. Moonlight Sonata lived up to its name. It is entirely possible, probable even, that the thirteen pathfinders of K.Gr. 100 would have found their target without the beams, that even if they had been jammed they would have dropped the incendiaries that started the blazing beacons on the horizon that drew the rest of the bombers in.

The problem is, we will never know.

In Berlin, the Nazi high command had no need of anguished counterfactuals. Joseph Goebbels was ecstatic, creating a new verb, 'Koventrieren' – 'to Coventrate' – meaning to reduce to rubble. Goering, too, was delighted. So delighted that he prepared a special New Year's message for the brave men of K.Gr. 100: 'I express to the C.O. and this Gruppe my sincere thanks for an achievement unique in history,' he told them a few weeks later. 'I know what enormous personal effort it has entailed on the part of each individual and I am convinced, my comrades, that in 1941 as well, you will know only victory.'

CHAPTER 9

Operation Starfish

'There is no place where a woman and her daughter can hide and be at peace. The war comes through the air, bombs drop in the night. Quiet people go out in the morning, and see air fleets passing overhead – dripping death – dripping death!'

H. G. WELLS, *THE WAR IN THE AIR*

November 1940

AT ABOUT THE SAME time that the crews of K.Gr. 100 were receiving a stirring message of support from Hermann Goering, Colonel John Turner was going through his own postbag. It, alas, made for rather less pleasant reading. The worst letters, he found, came from wealthy landowners. Or as he put it, in a document that spoke of weeks of accumulated frustration, from 'the pompous country house owner'. Often, he said, these 'egotists' would enlist 'self-appointed local busybodies' (a class that for him, stretching the definition of 'self-appointed', included Members of Parliament) to 'provide the most virulent and inaccurate diatribes and demands for disciplinary action against officers'.

It was a stressful time for Turner, and being assailed by MPs in the pocket of minor aristocrats made it understandably more stressful.

On the other hand, despite his unflattering descriptions of them, it is possible to sympathize ever so slightly with his correspondents. It was Turner's job, you see, to find fields on the approach to major cities, fields in and around those landowners' farms, and then, night after night, to set them on fire.

In these fields were increasingly elaborate devices, electrically controlled, to simulate different kinds of blaze. There were the 'basket fires', raised containers layered with creosote, pine shavings and scrap wood. These would produce intense conflagrations lasting up to an hour. The sort of fire, perhaps, that might also come from a burning storage depot. There were the 'boiling oil fires', in which oil was sprayed on to a coal fire, until it vaporized and burnt with vigour – at which point a water tank spilled its contents on to the already intense blaze, sending oil spitting and crackling. This created the sort of sight that might look like, say, a fuel dump going up. Between these were the coal fires. These produced dull, more prolonged blazes that would continue to glow for many hours. Similar, maybe, to the embers of a burnt-out warehouse. Next to them, there were sheets of asbestos. These weren't to keep the different fires under control – although that could be a problem. Instead, they were there to reflect the glow of the fires in much the same way as, say, a brick wall would.

From the field itself, it all looked bizarre. With angular, random braziers raised on stilts it was like some sort of modernist interpretation of Guy Fawkes Night. From above, though, perhaps with a bit of ground mist, a bit of luck and – ideally – a frightened and panicky observer eager to do his job and return home, it was different. Then Turner's fields, set up at a rate of one a day, became a passable impression of a burning city.

This, more than anything, was what must have troubled Turner's angry letter-writers. His goal – indeed the entire raison d'être of his war work – was that bombers would see the fires and assume they were over their target. Then, they would attack. He had an easy method for getting instant feedback on how effective he was: after a raid, he went and counted the bomb craters in and around his contraptions – for at least half a mile in all directions. If you happened to live within that half-mile radius you were, quite simply, bait.

Turner was in charge of what became known as Operation Star-fish. Operation Starfish in turn existed because Goering had a point when he sent his herogram to the pathfinders of K.Gr. 100. The group were indeed very successful, and it was far from clear – even with radio countermeasures – that they were going to cease being so. The British had found Knickebein relatively easy to jam, but X-Gerät, although superficially similar, was a different proposition. First, for the British to work out the X-Gerät target each night was a feat of technical skill. There were twenty possible frequencies that the Ger-mans could broadcast on, and the system was slick enough that it could be swung to a new target midway through the night: there was a lot of the spectrum for the British radio operators to search, a lot of sky to search, and not a lot of time in which to do it. Even if they did find the beams in time, the complexity of the radiation pattern from the many beams involved, Jones complained, made it harder to sepa-rate each out and find their routes and crossing points – the places that determined the target.

But for all the sophistication of X-Gerät, it also had a huge flaw. It had been rushed out too fast. During the summer and early autumn of 1940, when Knickebein was still in use, K.Gr. 100 had operated on its own, testing the system. In the right conditions, it worked excep-tionally well. Intercepted reports of a demonstration for Goering found that the bombs were landing between 100 and 300 metres from their target – unheard-of accuracy. British analysis of real Ger-man raids on targets in the country, where the bombs were dropped from 20,000 feet over hostile territory, found that on average they hit just 120 metres left or right of the main director beam. In terms of range, however, they were less accurate. They were landing close to the beam the bombers were flying down, but the bombs were still being released hundreds of metres too late or too soon. More than likely they flew over the target; but they didn't release over the target. Partly, Jones thought, this was because of 'slovenly setting of the clocks' that measured the bomber's speed through the air. If so, it could be rectified by the Germans. 'If the cross beam accuracy were improved (as could easily be done) the threat, in the absence of effective countermeasures, would be serious.'

Yet this improvement had not been made. Instead, with Knicke-bein faltering due to Jones's work, the Luftwaffe 'were forced to extemporize'. They desperately needed another way to guide their bomber force. 'In the search for a substitute they found that K.Gr. 100 was still getting to its target completely unmolested.' There was no time to produce and install enough X-Gerät receivers to equip the whole bomber force, so this was when K.Gr. 100, using a system not yet perfected, became the pathfinders for their colleagues.

The group was led by Kurt Aschenbrenner. His surname was also the name of an occupation – someone who works with fires and ash. That is what his crew were now going to do. They would, in Jones's words, 'blaze a nightly trail for the Luftwaffe'. In switching to the lighter incendiaries, which could be tossed on slipstreams and in the turbulence of clouds, they also lost much of the value of the system. 'It is obvious,' wrote Jones, 'that from the moment of applying the new policy all attempts to extract the utmost accuracy from the X-system were literally thrown to the winds.'

More than that, in this supremely automated system, they had once again introduced human fallibility. The main bomber force knew simply that they had to drop their bombs on the fires created by K.Gr. 100. What, though, if there was more than one fire? What, in fact, if the first fires they met – roaring blazes an acre or more across – were not set by the pathfinders at all? That was why Turner was setting fields alight. Starting in late November, and then at a rate that peaked at one site a day, he built Starfish sites across the country. He was Britain's pyrotechnician in chief.

Often, the decoys were damp squibs. The beam routes were deci-phered too late, the city itself burnt too well or the night was too clear – showing up Turner's fires for what they were. Sometimes, though, everything aligned. There were nights when the listening service found and plotted the beams in time. The emergency ser-vices, alerted to the target in advance, were on hand to douse the incendiaries as they landed. The Starfish oil fires on the city outskirts produced a crackling and spluttering extravaganza in time for the main force, and in blackout Britain, far from the safety of home, the bombers above thought they had found their target. Then, the combined technological might of the most technologically advanced force the world had ever seen wasted tens of thousands of Reichsmarks of high-tech high explosive in bothering some worms – and also, though obviously Turner would not have wished it, the adjoining fields of some pompous local busybodies.

Turner knew when he was successful. When, on 3 December 1940, he went to one of his Bristol sites he found sixty-six craters – which came from sixty-six bombs that had not landed on the streets and factories beyond.

Aileen Clayton, working in radio intelligence at the time, used to take great pleasure in hearing the planes calling home, to indicate a successful raid that was anything but. 'We were delighted when we heard a pilot reporting, "Target attacked . . . several fires blazing", when we knew his bombs had fallen inoffensively on a Yorkshire moor rather than the factories of Derby,' she said.

From POW interviews, it was clear that counting craters or listening out for radio intercepts measured just one benefit from the fires. As with the jamming of Knickebein, there were psychological consequences too – as word spread of the decoys, Jones reported 'subsequent diffidence of the Germans in bombing genuine fires.' Squinting at the bomb sites from 20,000 feet, were those glowing embers the remains of British factories – or long rectangular baskets of coal? Were the dull yellow flames reflecting off nearby walls, or asbestos sheets? The crews began to doubt their own eyes.

But if success was obvious, so too was failure.

On 28 November, Jones left the office knowing that an X-Gerät raid was on its way. This time, though, Bletchley had failed to decrypt the target in time. A few hours later, Jones was almost the first in Britain to find it out: he could see it from his window. 'We heard a noise that is unforgettable to anyone who has experienced it,' he wrote, 'sounding rather like ghosts in hollow chains rattling across the roof . . . I went over to the back door of the flat, from which we used to admire the view across to Windsor Castle.'

Everywhere they could see incendiaries. 'The sight was fantastic – a panorama stretching far away to Kew Gardens, with all the domes, spires and trees silhouetted in the pale blueish light.'

He knew exactly what it was: the X-Gerät pathfinders of K.Gr. 100. Soon, the proper bombs would fall. Vera, by this stage, was heavily pregnant. But, equally, it was quite the view. 'I called the others to the door, saying, "Come quickly, this is a sight you may never see again!" And then after a few spellbound seconds I said, "Now run downstairs like hell!"' For them, it was an exciting brush with the enemy – a night when danger was close, but not too close, and, in the communal air-raid shelter, they bonded amid their shared exhilaration. For many, danger came far closer.

Just as Turner had the nights when everything went right for him, so the Luftwaffe had their victories – the missions when all the stars aligned. Few more so than what became known as the Christmas Blitz on Manchester.

Around the middle of December, Mary Corrigan had been listening to the radio, to the broadcasts of Lord Haw-Haw – the Nazi propagandist. He said, 'The Manchester people have bought their turkeys for Christmas, but they won't cook them.' Then, the evening of 22 December came, she recalled, 'and the sirens went to let us know the air raid was starting.'

What came next could have been taught in the textbooks. At 6.41 p.m., the first bombs fell – and they were incendiaries. K.Gr. 100 were overhead, dropping the incendiaries that lit the path. Each weighed just a kilogram but, collectively, they lit up the city.

Dr Garfield Williams was Dean of Manchester Cathedral, and would never forget that night. 'The cathedral in its setting was a thing of entrancing, shocking, devastating beauty,' he said afterwards. 'I choose these descriptive words advisedly. All around, instead of hideous ugliness, there were flames shooting, apparently, hundreds of feet into the sky.' The wind, he said, was 'so filled with sparks as to give the effect of golden rain'.

That was only the start. The pathfinders had lit the fires to draw the main bomber force in. By 7 p.m., high explosives had taken over from incendiaries, and did not stop falling for eleven hours.

The very last bomb that fell that night was witnessed by Williams, who was right by it. 'It dropped on the north-east corner of the cathedral. The noise of the fires was so terrific that we did not hear anything. The sensation was just like an earthquake. The blast had lifted the whole lead roof of the cathedral up and then dropped it back, miraculously, in place. Every window and door had gone; chairs, ornaments, carpets, furnishings, had been just swept up into the air and dropped in heaps anywhere. The High Altar was just a heap of rubbish ten feet high. The two organs were scattered about in little bits. The Lady Chapel, the Ely Chapel and much of the regimental chapel had simply disappeared. Showers of sparks still swept across the place, but the old cathedral just refused to burn.'

Turner kept on with the Starfish sites, but they were never going to be enough. Where were the bromides? Were they working at all? Without physical evidence on the ground it was hard for Cockburn's team at the Telecommunications Research Establishment, who designed the bromide jammers, and Addison's at 80 Wing, who operated them, to judge their effectiveness. Were the dashes they sent into the ether causing at least some bombs to miss their target? Were their signals, hastily retuned each day, redirecting the beams, confusing the electronic equipment just as the aspirins had for Knickebein and its human listeners? Or were the stray bombs that fell off target, as many did, simply a reflection that bombing was hard? What was clear from the Manchester bombings was that however well they worked it was not well enough.

On the 23rd, after a night spent sitting in the cellar of her house, Mary Corrigan went up to see the dawn – and to return to work. 'There were no buses. There was no bomb damage near where I lived, but on the other side of the road, coming up from the city, was a long line of refugee people, who had been bombed out during the night and lost their homes. They were coming up and carrying whatever they could . . . a long line of refugees coming to Prestwich to see who would take them in.'

Throughout the day, rescue crews worked to save people. There had been too many acts of heroism to retell. The night had had its regulation dose of babies born in the back of ambulances, of daring rescues from collapsing buildings, of miraculous escapes of firemen. Then, at 7.18 p.m., even as exhausted fire crews worked to put out the last of the blazes, the air-raid sirens went again – and fire rained down once more. The next day it would be reported that the fire in Manchester's Piccadilly was the biggest seen in Britain since the Great Fire of London.

It would take a while for the radio experts at TRE to find any evidence at all that their efforts were not in vain – that their continued work to ramp up the electromagnetic screen, to increase the number of bromides, served any purpose at all. One clue came on the night one of the bromides failed. There was a raid in the West Country where it was possible to jam the cross beams but not the approach beam, a report explained. On this occasion the bombs were 'extremely accurate in line, but had an error of one mile in range.' This was interesting but also, as other analyses had shown, not definitive. The accuracy of X-Gerät relied on flying straight, level and at a constant speed for the last 20 kilometres or more. Get that wrong and you would still be on the beam – and so be extremely accurate in line – but completely out on range, with or without jammers. Another clue came, yet again, from POWs. A K.Gr. 100 crew crashed in March 1941, and were interrogated. They talked about how the beams' usefulness had 'deteriorated seriously' over the preceding months. In December and January, they said the interference to the approach beam was 'very serious'.

The most sincere compliment the Luftwaffe paid to the work of the bromides, and the surest sign they had an effect, though, came in their beam operators' attempts to evade them. Where once the beams had been switched on and made ready in advance of the nightly raid, by the New Year it was becoming standard to shift and switch frequencies at the last moment. As K.Gr. 100 approached the targets, suddenly the beam would appear to those jamming it below to have been turned off. On board, the operator knew to switch to a different Anna number. Below, without knowledge of what that number was, the jamming became useless.

It was a sensible precaution, but also a crude counter-countermeasure. For 80 Wing, in this endless arms race of electromagnetic one-upmanship, the counter-counter-countermeasure was simply to become more efficient at what they did. Rather than reporting leisurely on the frequencies in use in early evening, then using the same ones all night, listening stations remained on hand for the switch – waiting by specially installed landlines to report back the characteristics of the new beam. Then, as soon as 80 Wing HQ received the new frequency, they sent it out to the bromides. 'It was usually possible to follow any such changes in the frequency within three or four minutes,' said Jones.

So it was that, through Britain's darkest winter, K.Gr. 100 was denied the use of the electromagnetic spectrum. Or that, at least, was the accepted narrative – the one believed by Jones and his superiors. But attributing causality amid the confusion of war is hard. It is harder still when you are dealing with untested technology on one side – and its interaction with yet more untested technology on the other side.

Alfred Price was an electronics officer in the RAF in the 1950s and, on retiring from that job, became one of the great post-war chroniclers of the aerial information war. In the 1990s, he was invited to a reunion of K.Gr. 100, and wanted to find out about their understanding of what, then, was known as electronic combat. He was surprised by the answer. 'I asked one radio operator how much trouble he had had from the British jamming of X-Gerät. He said he did not remember any jamming. I described the Morse-dash jamming put out by Bromide, but he said he had never heard any such jamming. He called over some friends and soon I had half a dozen ex-radio operators standing around me and discussing the matter. None of them remembered hearing the British Morse dashes.' Of course these recollections, of ageing crews half a century on, directly contradicted those of some crews captured at the time. Equally, though, here was strong evidence that bromides on their own did not win the war against X-Gerät.

Price's bet was that the reason K.Gr. 100 never managed to 'Coventrate' Britain was less to do with one single countermeasure than all combined – and more. It was to do with the fact that the Knickebein countermeasures forced its early use, in just one small group of bombers, a group that itself became depleted and exhausted. It was because the pilots themselves were affronted by the system. General Wolfgang Martini said in post-war interrogation that those in the main bomber force considered relying on pathfinders to be insulting. 'The pilots maintained that they could obtain better results by finding the targets themselves, and those who were obliged to follow the pathfinders felt that they were being relegated to subsidiary tasks. They reported that K.Gr. 100 had bombed in the wrong place, and that they had found the right target by themselves. Despite the good results obtained with "X", anger against the system remained unabated.'

Then there was also the most important wartime factor of all to consider: sheer luck. With Coventry, everything aligned in a way it could not elsewhere. 'It was the rare combination of perfect weather, a full moon, a highly combustible target and weak defences that sealed the fate of Coventry,' Price suggested.

Whatever the reason, as the winter progressed Jones felt confident to say that Goering's enthusiasm for his pathfinders in K.Gr. 100 had diminished. It became obvious to planners that the enemy were falling back on more tried and tested navigation methods. Targets were increasingly those that could be found using natural features such as the Thames, coastlines, or even, Jones wrote, a source of reflected electromagnetic radiation that even TRE could not jam: the moon.

'By April, the bulk of the Luftwaffe was generally being employed on moonlit nights, showing a better precision than had been attained with fire raising,' he said. Although, he added, this precision came at a cost – in seeing the ground better, they too could be seen better, 'providing better sport for our night fighters'.

Across the UK there was a growing defiance, and a sense that the worst was over. As the first leaves appeared on the first springtime trees, a new national myth was already being formed: of resilience, of keeping calm and carrying on, of strength forged in adversity. Less than a week after Manchester's Christmas Blitz, a correspondent for the *Manchester City News* reported from the centre of the city, the rubble still not cleared.

Instinctively, one recognized the calibre of these people. An intense sense of unity has been created. If Hitler visualized these folk as shrieking and tearing their hair and cowering before his rage he had another guess coming.

When I penetrated to where the rivers of water ran through shattered glass and snuffed the tang of smoke and saw little flames flicker onto the skeletons of buildings ... and grimy and weary firemen and ruddy-cheeked soldiers with fixed bayonets ... I knew that I loved Manchester. Its dear smoky streets, its kindly, comradely folk, the very nooks and alleys of it – I loved it.

If this be the battle of Manchester then Hitler has lost it.

CHAPTER 10

The King of the Gods

'With our enormous metropolis here, the greatest target
in the world, a kind of tremendous, fat valuable cow, tied
up to attract the beast of prey, we are in a position in
which we have never been before, and in which no other
country is at the present time'

WINSTON CHURCHILL, SPEECH TO THE HOUSE OF
COMMONS, 30 JULY 1934

Christmas, 1940

IF YOU WERE TO paint a picture of a traditional English city pub, it
would be something like the Feathers, in Broadway near St James's
Park in central London. Just below the first floor – brick, neat and
terraced – are two cast-iron gas lamps. They illuminate its wooden
sign, the name commemorating the crest of the Prince of Wales.
Below that sign in turn is the sort of ornate Victorian marble and
glasswork facade that is just the right side of fussy. This is the kind of
pub where you expect a big shaggy dog asleep in front of a roaring
fire, a well-built landlady who can defuse a fight just by folding her
arms and glaring, and to find, at any time of the day, a group of comfortable regulars and a badly tuned piano. And if, towards the end of

1940, you had entered it to see a large crowd, elbows jostling at the bar, swaying to a harmonica sing-along? Well, that would have felt just right too.

With the festive season had come, unexpectedly, something of a hiatus. Each afternoon, it had become Jones's and Frank's routine to check on the beam settings. When combined with Enigma intercepts and analysis of signals traffic going to and from airfields, Knickebein and X-Gerät would give them an indication of what Goering was planning.

During the afternoon of Christmas Eve, for the first time since 7 September, there was radio silence. 'It became clear,' Jones later recalled, 'that the German Air Force intended to celebrate the occasion in traditional style by not coming out to attack us.' The wife of one of his colleagues came late that afternoon to pick up her husband, and happened upon Jones instead. He passed on the good news – they could have a peaceful night's sleep at home in their own bed for once. She was unimpressed. 'To hell with our going home,' she said to Jones. 'What's more, you're not going home either – you're coming to the Feathers to have a drink with us.'

For the employees of the Minimax Fire Extinguisher Company, the Feathers had become their local. Or, as Jones put it, it 'tended to serve as an annexe of MI6 headquarters'. On that particular evening, they arrived to find that Britain's foreign intelligence service had decamped *en masse* and was 'doing their best to observe the Christmas spirit'. It did not take many drinks for Jones to be persuaded of the imperative for a collective sing-along, and he took his harmonica out of his pocket. As he played, his colleague's wife – also feeling convivial – picked up her husband's RAF cap and began passing it round. 'To my embarrassed gratification, they collected 51s 6d. I salved my embarrassment by giving the money to a Spitfire fund,' he recalled.

He also recalled the tale later relayed to him by a squadron leader who, in the pub at the same time, had handed the cap to a civilian punter in a somewhat 'happy state'. 'Without our inside knowledge about the German Air Force he had clearly decided that bombing or no bombing no Germans were going to stop him celebrating

Christmas Eve. He promptly threw in half a crown and asked the squadron leader, "Who's the chap who's playing? What does he do?" The squadron leader replied, "I don't know, but he's a scientist of some sort!" Whereupon the man put his hand into his pocket, pulled out another half crown and dropped it into the cap, saying, "They're badly paid, poor buggers!"'

Few could have begrudged them the party. Fewer still – even those from MI6 – would have known just how much the country already owed to the man playing the harmonica, a man still in his twenties. Thanks almost entirely to Jones, the UK had identified not one navigational beam but two. In both cases, it had done so almost before they were used – and in time to introduce the countermeasures that helped defeat them. It was only a brief respite, though.

Christmas is the celebration of Christ's birth. It is also something older. It is a midwinter festival, an ancient celebration of the return of the Sun. At this time of holly and ivy, of wise men and stables, the pagan has always mixed with the Christian – there has always been a link to something deeper in all our prehistory. So it was too in 1940, when even as churches prepared for Christmas Day ceremonies without bells, as mothers went to pray for their sons in North Africa and beyond, Jones's mind was on a more primitive, more earthy deity – one who had come to his attention months earlier and since then had been a constant niggle.

Wotan was the king of the gods and the father of Thor. From Valhalla, he presided over the living and the dead – assisted by his band of warrior maidens, the Valkyries. He was all powerful and he was always wise. In the beginning he made the earth and the sky, the first woman and the first man.

This was not what came to the mind of an academic called Frederick Norman, when Jones called him up one day many eventful months earlier in the summer of 1940 to ask about a curious message. Jones at the time – before X-Gerät was a concern – was extremely interested in anything to do with Knickebein. Reading through one decrypt, though, it was another word that caught his eye. 'It is proposed to set up Knickebein and "Wotan" installations near Cherbourg and Brest,' it said. What, Jones asked Norman, could it mean? What linked Knickebein and Wotan?

In peacetime, Norman had been Professor of German at King's College, London. In 1940, he was one of the resident linguists at Bletchley Park. Jones later recalled this conversation as a eureka moment. 'He said, "Well, he was head of the German Gods." Then he said, "Wait a minute, one eye!" Wotan only had one eye. One beam! Can you make a system work with one beam?' Jones realized they could. What's more, he realized such a system could at last be the one described in the Oslo Report.

The document provided to the embassy in Oslo less than a year earlier did not actually talk of beams at all. Instead, it described a rangefinder – something that told you exactly how far you were from a particular radio beacon.

Imagine sending out a radio signal in all directions, and imagine that that signal is a rapidly undulating wave. This wave travels unmolested through the sky, leaving the original beacon at the speed of light. Somewhere in its journey, a fraction of a second later, it meets a plane equipped to receive it – a plane with this rangefinder. This plane spots the wave, recognizes it, and then – here is the clever bit – instantly re-radiates it back exactly as it found it, sending it back to the first beacon. To the observers at that beacon, it will appear that the instant they sent out their signal they saw it return. But if a human can't spot the delay, electronics can. Because the returning wave will be ever so slightly out of synchrony.

The signal the beacon is sending out undulates. It undulates because that's what radio waves do – vibrating superfast from crest to trough. But it also undulates because, embedded in the signal, the Germans had put a more languorous wave – that went from crest to trough and back again a mere 3,000 times a second. This was slow enough that the time taken for a single wave represented light travelling 1,000 kilometres. When the beacon broadcasts a crest, by the time it has travelled to the plane and then been re-radiated back, the time delay means it could be sending a trough instead.

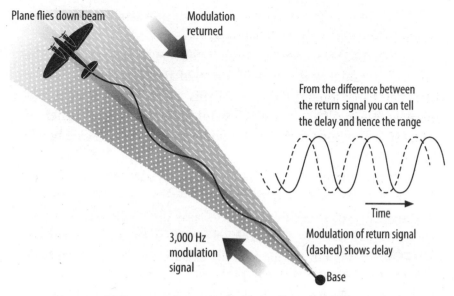

Plane flies down beam

Modulation returned

From the difference between the return signal you can tell the delay and hence the range

Time

3,000 Hz modulation signal

Modulation of return signal (dashed) shows delay

Base

Y-Gerät used a range finder and single beam

From this 'phase difference', the extent to which the two waves differ, you can tell how far the signal has gone on its round trip – so long as it has not gone more than 1,000 kilometres. The radio operator on the plane needs to do nothing at all. He simply waits for the controllers on the ground to subtract one wave from the other and then tell him how far away they are.

This system, on its own, is of marginal use. It's handy to know you are, say, 200km from Cherbourg, but that still leaves you anywhere on a circle encompassing Paris, Nantes and Bristol. If the same beacon also projects a beam, though, a straight line that, like a clock's hand, intersects the circle – well then, suddenly you have a single coordinate. Without needing cross beams or warning beams you can know exactly where you are at all times.

Could this be Wotan? A single beam, with an in-built rangefinder? In the months that followed, Jones was busy. His department of one was occupied with countering the already established Knickebein beams, and then with parrying the new upstart X-Gerät. But, throughout, he never forgot about Wotan. In July he filed a preliminary report, mixing supposition with the scanty knowledge in his possession to sketch out how the beam might work.

It would have been so easy to ignore Wotan, or to elide it with the other beams. Just as a lost hiker can convince themselves that the landmarks they see in front of them are those on the map in their pocket, even when they are not, the temptation in intelligence is to fit new facts to old deductions – to make square pegs (or, even, oval pegs) fit in round holes.

Jones did not fall into this trap. Just as with X-Gerät, slowly at first, then with more clarity, he was able to fill in the details of what, to him, was clearly a separate system. In November, an Enigma intercept came through. It contained the instructions the Germans had given to a particular beam station. Ordinarily, you would expect them to tell the beam where to point – to give it a direction. Instead, they gave it coordinates: a single spot on a map. What was also odd was that it was the only one to be told to point that way. Normally, with X-Gerät, Jones would have expected other beams – the cross beams that determined the target. There were none. Somehow, this one beam station was enough to provide a plane with all the information on its own to hit a target.

Then, one night just before Christmas, a radio listening station heard something strange. Over Southend, a Luftwaffe aircraft was being given instructions. 'Turn round and make a new approach,' the German controller was heard to say. 'Measurement impossible, carry out task on your own.' There was a blast of dashes on a particular wavelength. It wasn't much, but it was enough. It was as if the two, the radio controller and the plane, were engaged in an elaborate dance – one guiding the other – but both were still trying to learn the steps. Wotan had at last turned his single unblinking eye on Britain.

Jones was, as it happened, completely wrong. The Germans had not named the system Wotan because it had one beam. Technically they hadn't named it Wotan at all. They had named it Wotan 2. And the Wotan that it was a sequel to? In a fact entirely missed until this point by intelligence, Wotan 1 was this beam's predecessor. It was the codename for X-Gerät, a system that used not one beam but five. Yet it didn't matter. Because while his reasoning was completely wrong, his deduction was, entirely coincidentally, entirely correct. Wotan 2 – also known as Y-Gerät – was indeed a system based on one beam.

If the Germans did make an elementary error, it was not in how they named Y-Gerät, but in how they used it. From the summer of 1940 until the spring of 1941, in a period when Knickebein failed and X-Gerät limped on, there were a small number of planes using Y-Gerät instead. Night after night, in the skies above Britain, the clues were there – waiting to be found.

And found they were. Sometimes the listening organization, that had now moved on from merely sticking ham radio enthusiasts in sheds on poles, picked up snippets of chatter and unusual radio blasts. Sometimes, they picked up the clear to and fro of pilots and ground stations trying it out. Sometimes Bletchley, which was now cracking Enigma with regularity, picked up messages that filled in more details of the system – smoothing off the rough edges of Jones's supposition. Analysis of known Y-Gerät raids showed the bombs to have fallen with 'abnormal accuracy'. Prisoners were lyrical about its advantages. One was overheard describing a time when a Y-Gerät-equipped plane flew through the condensation trail of the craft ahead.

Sometimes, the answers came literally fluttering down from the sky above. On 19 January a Heinkel III was shot down near Eastleigh just before 9 p.m. All the crew were killed, much of the equipment was mangled, but a charred and battered notebook was found. It had three tables. The first, labelled 'return flight', detailed the distances to the plane's home airfield from five cities: London, Bristol, Sheffield, Southampton and Birmingham. The second, for the outward flight, detailed the bearing to take to reach those cities, from a point it was easy to determine was Cassel, a French town on a prominent hill south of Dunkirk.

From this, Jones made two key deductions:

(a) The aircraft approached its targets from the direction of Cassel.

(b) It was not concerned with distance calculations on the outward flight, and therefore employed some special method of calculation.

If the first two tables described the system in theoretical terms, the third provided the nitty-gritty:

Target Beam, 43.5 mc/s

Target Rangefinding: Receive, 42.3 mc/s, Transmit, 47.4 mc/s

The radio operator had done the 1941 equivalent of pinning a post-it note of his banking password on to his screen. Here was the key to Y-Gerät. It told him that pilots flew down a beam that they searched for on the frequency 43.5 megacycles per second. They received their automatic rangefinding signal on the nearby frequency 42.3 mc/s. They re-radiated it back at 47.4 mc/s.

The flaw in any apparently secure system is almost always people, and here was yet another proof of that adage. It was not merely the dead radio operator who had been sloppy, though. It was the air force that put him there. Why was Y-Gerät in use at all? Why reveal your hand before you need to? Here, again, it seems the answer lay in human psychology rather than military strategy.

What was interesting about Y-Gerät was that it was so very different from X-Gerät. Even its director beam, while relying on a similar principle to guide the bombers to their targets, in practice came at the solution in a distinct, almost divergent fashion. It was as if the two teams had had no interactions during development. Or, an alternative interpretation, it was as if Y-Gerät was purposely, almost perversely, staking its independence. 'These changes suggested tension of an original mind seeking to produce a system which, while mature in concept was nevertheless to have as little as possible in common with the earlier system,' Jones wrote. 'It is reminiscent of one commercial firm trying to circumvent the patent rights of the other.'

Very reminiscent. Johannes Plendl, the radar pioneer, was the creator of Y-Gerät and before the war had been employed by Telefunken. Rudolf Kühnhold, his protégé, developed X-Gerät, and had worked for Siemens-Lorenz. There was, Jones suspected, a power struggle. 'Rivalry between the two men has frequently been evident,' he said. 'Although Kuehnhold is considered Plendl's junior, the operational importance of X-Gerät has enabled him to meet Plendl on terms of equality.' There, he could see through intercepts that they fought not merely for status, but also for resources. They used the same transmitters – and for Plendl to take one for his system meant depriving Kühnhold of one for X-Gerät. In the post-war assessment of the air war, British intelligence later gloated over the way the Luftwaffe was 'beset by quarrels in high places' over the systems – leading to clear strategic mistakes among the leadership and 'further quarrels and uncertainty' among operational units.

'The squabble went on through September and October, but at the end of November, the director of Germany signals ruled the particular generator must be handed over to Dr Plendl,' wrote Jones. 'We had been waiting for such a decision as an indication of the attitude of the German Air Staff: it had come down on the side of Wotan.'

From Plendl's point of view, as he faced the upstart Kühnhold, it was a triumph. Thanks to the apparent promise of Y-Gerät, Goering made him a 'Staatsrat', a state councillor. It was a civil honour, given by Goering in one of his other capacities – as Prime Minister of Prussia. It was, Jones hoped and expected, premature. 'We may yet hear of Goering assuming another of his pooh-bah roles in order to reprimand his new Staatsrat.'

From a military viewpoint, though, he said, 'the decision was in all respects bad'. Bad, that is, for the Germans. Quite how bad would become apparent as the first signs of spring appeared.

A decent chess player confounds the opponent, countering each attack with a deft block that renders it useless. A good chess player stops the attacks before they have begun. A great chess player prevents the attack without his or her opponent realizing – until they are already committed. Ever since the Knickebein chase, Jones had a dream to not just beat the beams, but get ahead of them – to be ready before they were even used.

The ideal countermeasure, he argued, went through three distinct phases. The first phase is when it is so subtle it is almost imperceptible, when the enemy thinks the correct targets are still being hit even when they are not. This phase inevitably cannot last long before discovery, which leads to the second phase. This is when the slow realization comes that all is not right. In this phase, the goal is to cause such confusion that even when the system is working perfectly no one believes it. The countermeasure, wrote Jones, needs 'to induce a mistrust in his pilots, so that they are not sure whether to believe their instruments and/or their senses, and thus jeopardize their efficiency.' Only then comes the final phase. Realizing that, from the start, they had been fooled without knowing it, the German staff – at least in Jones's imagining – would throw in the whole beam contest entirely.

For X-Gerät and Knickebein, there was no time for such luxuries: crude jamming was needed immediately. For Y-Gerät, perhaps thanks to Plendl, there was space to think and room to be clever.

Six years earlier, the BBC had first been called to do its duty in the radio war. A creaky Haverford bomber had flown past a short-wave transmitter, and the faintest of reflections had come bouncing back – a radar echo. Now, the BBC was to be called on once again. If the first mistake of the Y-Gerät team was to reveal its hand too soon, the second was to do so on a frequency that happened to coincide with that used by one of Britain's most powerful transmitters: the mast at Alexandra Palace.

When the Y-Gerät pathfinders flew over Britain, their rangefinding signal came at 42.3 mc/s and was retransmitted at 47.4 mc/s. It was pure bad luck for the Germans that when the early adopters of television turned the dial on their machine, waited for the cathode ray tube to warm up, and then squinted to see the output of the 'BBC Television Service', the black-and-white images they received also came on a frequency between 40 and 50 mc/s.

Except that in 1941 they didn't. The early adopters of John Logie Baird's invention – an invention memorably described by an incredulous news editor at the *Daily Express* as 'a machine for seeing by wireless!'* – were not seeing a great return on their investment. On 1 September 1939, the BBC broadcast *Mickey's Gala Premiere*, a Mickey Mouse cartoon. Then at 12.15 p.m., the show ended with Mickey in an embrace and the words 'I kiss you now', and the service went off air. Fearing that the Luftwaffe could use the transmitter as a beacon, and with almost no one watching TV anyway, the decision had been taken that it was to be switched off for the duration of the war. Or, rather, that was the official story.

In the winter of 1941, the BBC's engineers received a message. The mothballed tower was to be reactivated, and put to work on a task even more important than letting the wealthy technophiles of London watch Disney cartoons at home. It was, as it happened, going to be a Luftwaffe beacon after all.

Jones described the plan as like feedback on a microphone. We have all seen – or, rather, heard – the wail of a microphone too close to a speaker. Perhaps it is someone giving a speech at a wedding, already nervous, pacing the stage. Perhaps it is a band, setting up for a pub gig. But if you bring a microphone in front of its amplifier there is a painful screech. The sound picked up by the microphone is sent to the speaker, where it enters the microphone and is sent to the speaker again – looping and looping at the speed of sound until it is a monstrous electronic scream, and audio short circuit.

Jones and his colleagues at TRE realized that they could do the same, but subtly at first. At Alexandra Palace, they would listen out for the return ping sent by the bombers – and then send it back to them. The German controllers would send out the signal, which would be received on one frequency then re-radiated back on the other. A fraction of a second before it returned to home, though, it

---

* Logie Baird had turned up at the paper to show off his machine – but the editor was suspicious of the story. 'For God's sake, go down to reception and get rid of a lunatic who's down there,' he told a reporter. 'Watch him – he may have a razor on him.'

would pass Alexandra Palace. Here, the engineers were waiting. When they heard this ping, they re-radiated it a third time, but on this occasion on the original frequency. It, in turn, was picked up by the bombers and returned – in an infinite loop that, like the feedback on a speaker, rendered the system useless.

It was just as Jones had hoped. 'The existence of deliberate countermeasures was not immediately obvious to the enemy,' wrote a highly satisfied Cockburn. Listening in to their back and forth with their home base gave a much-needed boost to morale among the nation's overworked radio scientists. 'Mutual distrust between the ground station and the aircraft existed for a considerable time. The internecine strife which occasionally developed was most heartening to the jamming teams.'

Early in 1941, the greatest of the Germanic gods turned his powerful stare into central England in earnest. Each night, the planes took off assuming the teething troubles would be resolved. Each night, they got slightly worse. The signal was turned up, the countermeasures improved, the pings of the planes re-radiated, interfering with the precise equipment. Jones wanted puzzlement. He wanted brainpower to be expended. He wanted the realization to come slowly – as indeed it did. He wanted the Germans to understand they were playing a grandmaster, and that they might as well give up. By the spring of 1941, with the third of their three beams countered, that is largely what they did.

There would not be a fourth beam. The great eye of the Nazi war machine was turning elsewhere, turning east. Soon, Britain would be not on the defensive but on the attack. And the electromagnetic war was going mainstream.

During those desperate years, the British public had believed they were protected, by anti-aircraft guns on the ground and nightfighters in the air. In truth, they weren't. In truth, said Jones, tracking, confounding and stopping the beams had been a key defence. 'Night fighters, until airborne radar was good enough, were powerless, anti-aircraft guns were inaccurate, therefore we could not stop the bombers coming. Our only hope was to throw them off. Well we didn't always succeed but we did on a fair number of occasions, with the result that a good many people were alive at the end of the war, who otherwise wouldn't have been, and a good deal less vital damage was done.'

Now, it was time to move on. The Battle of the Beams was over, but Jones's war was not. Neither was the battle for electromagnetic supremacy. Far from it. In fact, it was going to get bigger. The battle Jones started had been joined, and became so big that this book can, like the electromagnetic spectrum itself, only illuminate a small sliver.

In later years, when Jones published his autobiography, contemporaries would claim that he took too much credit for his role. Quieter scientists, they argued, were forgotten – crucial contributions overlooked. It is true that many others were involved, particularly in the pages to come – and I will tell their stories where I can. It is also true, though, that for those first two crucial years Jones, like Britain itself, often found himself standing alone.

After the war, Charles Frank was asked by a sceptical scientist if Jones really was as crucial as he claimed. 'Was he really as good as he makes out, you ask?' Frank replied. 'We were bloody good at our job then, and we worked with some bloody good colleagues; and in our particular subject area . . . we were better than anyone else who tried to muscle in; and,' he continued, 'for the first fourteen months "we" was just R. V. Jones.'

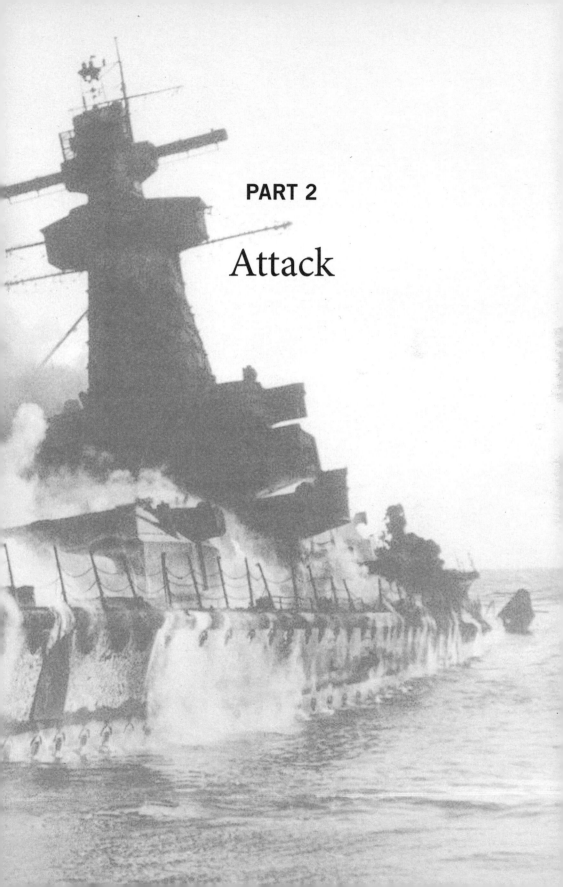

PART 2

Attack

CHAPTER 11

On a Plate

'Better a thousand live young men than a thousand
dead heroes'

CAPTAIN LANGSDORFF,
ON HIS DECISION TO SURRENDER

December 1939

CAPTAIN HANS LANGSDORFF SAT in his Buenos Aires hotel
room, a broken man. At the Hotel Centro Naval he could, if he
had wanted to – and he probably didn't – have walked to a window
and seen out to the estuary of the River Plate beyond. This river is the
confluence of two others that have meandered for thousands of miles
through the jungle, canyons and mountains of South America.

Opposite his hotel, their journey ends – widening into a powerful
channel that marks the border between Uruguay and Argentina. At
that particular moment, sitting half submerged in this estuary, its
hull stuck on a mudbank, lay the listing and smoking hulk of his ship,
the *Graf Spee*.

The newspapers in Argentina were already mocking Langsdorff.
One called him a 'coward and a traitor to the tradition of the sea'
because he hadn't gone down with his ship. The truth was even worse.

His ship hadn't even gone down – but it should have done. On its final voyage, a voyage he and a skeleton crew had undertaken to destroy it, he had not been able to reach deep enough water to sink it, and prevent it falling into enemy hands. Instead, there it was – for all to see.

Langsdorff couldn't even scuttle his ship properly. He went to his desk and wrote three letters. The first was to the German government, to be passed on by the ambassador. The second was to his parents. The third to his wife.

Just a few days earlier, the prospects for the *Graf Spee* had seemed so much sunnier. For weeks, the battleship had been harassing shipping in the South Atlantic and in the Indian Ocean. She had sunk eleven merchant vessels and, following German naval policy of the time, avoided any engagements with enemy warships. There was less glory in sinking unarmed ships – but a lot more tactical value. At the start of December, the inevitable happened – and she faced the prospect of battling her naval equals. Seven 'hunting groups' had been organized by the Allies – totalling twenty-three ships – and tasked with seeking her out. There are a lot of places to hide in the Southern Atlantic if you want to. There are fewer, though, if you also want to keep on sinking ships.

The Allied hunting groups had set sail for the most likely targets. One of those was the point where the River Plate meets the ocean. Here, at 6 a.m. on 13 December, Hunting Group G, consisting of the three cruisers HMS *Ajax*, HMS *Achilles* and HMS *Exeter*, were steaming in line ahead when they spotted smoke on the horizon. It was, at last, the *Graf Spee*. And she had seen them too. Minutes later, she opened fire. It was three against one but, as the next three desperate hours showed, that did not mean it was an unfair fight. What became known as the Battle of the River Plate was short, brutal and – at the time – inconclusive.

Commodore Henry Harwood, in charge of Hunting Group G, split his ships. *Exeter*, the heaviest, swerved to pass on one side of the *Graf Spee*. *Ajax* and *Achilles* went for the other. In this way, he hoped to force the *Graf Spee* to choose which to attack. She chose *Exeter*, and her fire was devastating. Hit after hit landed. The captain was wounded in both eyes and both legs. Gun turrets were disabled one after the other. As her last gun turret failed, the *Exeter* was forced to retreat for repairs.

It was not all one-sided. The *Graf Spee* was hit too. One shell nearly took out her control tower – lodging splinters in Langsdorff's shoulder. Another briefly knocked him out. The simple fact was, like heavyweight boxers slugging it out, even when it was three on one the German ship could take more punishment. The *Ajax* and *Achilles* pulled back to shadow the *Graf Spee*. To them, it looked like she was as seaworthy as ever; but deep beneath the hull, as she steamed through the South Atlantic, all was not well. Of the twenty-three Royal Naval shells that hit their target, one had been a very lucky hit indeed. Passing through two decks, it had destroyed the steam plant. This was used to make fresh water, and also to power a turbine that filtered out pure diesel from the oil that the ship carried. Without it, the *Graf Spee* could not get home. There were just sixteen hours' worth of usable fuel left.

Langsdorff had no choice but to head for the nearest port: Montevideo, in neutral Uruguay. Out at sea, the *Ajax* and *Achilles* patrolled menacingly, shadows on the horizon. In port, other Allied forces were gathering, as a naval battle became a diplomatic one.

Sir Eugene Millington-Drake, the British ambassador in Montevideo, had a difficult task. The *Graf Spee* could not be allowed to stay in port long enough to do repairs. Arguing that the ship was essentially seaworthy, he lobbied the Uruguayans to give it twenty-four hours. They went for forty-eight. But after this triumph, the Navy then asked him to do the reverse. They still didn't want it to have time for repairs, but they also realized they couldn't let it leave just yet. They still had just two cruisers blocking the escape, and it wasn't enough. They didn't want the *Graf Spee* to sail out before reinforcements had arrived. So now a delay was needed. Here, Millington-Drake applied an arcane bit of international law. He convinced a British merchant ship to sail out just before the 48-hour deadline. Under the rules of the Hague Convention governing the behaviour of combatants in a neutral port, the *Graf Spee* had to give it a day's head start.

During this hiatus, Langsdorff was growing increasingly worried.

In the port's bars, rumours began to fly that this was only the beginning, that a larger British force had arrived, a force that would ensure the *Graf Spee* could never break out. If anyone could have traced the rumours to their source – and a lot of work went on ensuring they couldn't – they would have found that they began with Millington-Drake's agents. On the third day, Millington-Drake put a call in to the Argentine authorities to bring in fuel for the expected arrival of two battlecruisers. There were no cruisers, but the Germans were listening in, as he knew they would be.

On the *Graf Spee*, the crew were getting increasingly jittery. The gunnery officer, peering at the horizon, claimed to have seen the battlecruiser HMS *Renown*, and then, later, the aircraft carrier *Ark Royal* accompanied by three destroyers out in the sea beyond. Langsdorff looked at the distant angular shadows of the Royal Navy on the horizon, guarding his exit, and after much agonizing accepted defeat. Placing charges on the hull of the *Graf Spee*, he tried to scuttle her, but in water too shallow for her to sink. For several hours, the flames from the ship lit up the coastline of Montevideo, while the crew crossed to the other side of the River Plate – taking refuge in Argentina, a country more sympathetic to the Nazi cause.

Pictures of the defeated ship went round the world. In the few months of the war so far, the Battle of the River Plate had proved an exciting diversion. Small, self-contained and dramatic, it had kept the British public rapt as they followed its progress.

Churchill immediately exploited it for propaganda. 'This brilliant sea fight takes its place in our naval annals and in a long, cold, dark winter it warmed the cockles of the British hearts,' he declared, adding that it would 'long be told in song and story'. As he probably knew, its morale value exceeded its strategic value. But if only the Admiralty had paid more attention to that ship, it might have been the other way round.

Washed in the waters of the River Plate, wedged at an angle on the mudbank, lay the blackened hull of the *Graf Spee*. As the smoke cleared, above its bridge an odd array of criss-crossing wires was clearly visible.

*

Before the war, there had been an uneasy detente between the Luft-waffe and the RAF. Delegations from each side would travel to the other, for visits that were ostensibly about building bonds of friend-ship, but in reality a game of bluff and double bluff – of revealing strengths and hiding weaknesses, of choosing which secrets to pro-tect and which to use to intimidate.

On one trip to the RAF's Air Staff College in 1937, two years before war broke out, the Germans were shown 'shadow factories', designed to increase aircraft production capacity rapidly. They were not shown the prototype Spitfire. On a reciprocal visit to Nuremberg, to coin-cide with the rallies, Frederick Winterbotham, the RAF officer who helped coordinate Enigma intelligence, asked how they kept the Hit-ler Youth under control. 'One teaspoonful of bicarbonate of soda per head per day, and sex doesn't even start to rear its perverted head,' came the surprising reply.

More crucially, also in 1937, Luftwaffe General Erhard Milch vis-ited Fighter Command Headquarters. In the officers' mess, he decided on an unusual conversational gambit. 'How are you getting on with your experiments in the detection by radio of aircraft approaching your shores?' he asked. Ploughing on, he continued, 'We have known for some time that you were developing a system of radio detection, and so are we, and we think we are ahead of you!'

The point of this story is that it should not have been a surprise that the Germans had radar. The Germans had literally told them that they had radar. And in the *Graf Spee* was the proof. Short of hav-ing Milch also hand over the blueprints, the *Graf Spee* was the next best thing. Even so, it took more than a week for anyone to notice.

In January 1940, an analyst looking at press photographs of the *Graf Spee* spotted the wires – an 'aerial' still attached to the bridge. Labouchere Hillyard Bainbridge-Bell, a radar scientist at the Air Ministry, was sent to Paris, where the Deuxième Bureau, the French secret service, had some better pictures. From there, he was suffi-ciently intrigued to make the journey to Uruguay. He took with him on the plane a newly published memoir entitled *I Was Graf Spee's Prisoner*. Just months earlier, Captain Patrick Dove had been cap-tured and imprisoned on the ship By the standards of later war memoirs, the story was tame. His was no dramatic escape from a prison camp or daring desert raid – he just spent a few weeks being well treated on a boat. But the war had barely begun, and the nation needed heroes, particularly those connected to the now famous *Graf Spee*. So his tale was written up and rushed through publication.

What stood out for most readers were the passages recounting his conversations with Langsdorff, and his meals with the crew. However, this was not what Bainbridge-Bell was interested in. In an aside, Dove describes a turret at the top of the ship that rotated and 'was manned and operated day and night. It never stopped.' Bainbridge-Bell was relieved on landing in Montevideo to see the turret still there. It might not have been. Souvenir hunters had gutted the lower decks. Nazi sympathizers, he feared, had hidden much of the rest – one would later steal some of the equipment he had salvaged, planning to take it home. Then there had been some wrangling and threats from the Uruguayan authorities. 'Fortunately,' wrote Bainbridge-Bell, wryly, 'the ship was not blown up.'

This was now his asset. British intelligence had, using a front company, purchased the scrap metal rights to the *Graf Spee*. Bainbridge-Bell was their scrap dealer. He rowed out to see it.

The brief document that resulted from his few hours on board is a masterpiece of analysis. In meticulous detail, he pieces together the intricacies of the radio system. From his position on a listing deck, ravaged by fire, he found the fragments of cathode ray tube that told him it had a display, and the scraps of electronics, gears and aerial that showed him how it worked.

'It is suggested,' he wrote, 'that the unit was used as a variable delay device. If this is so, the ranging operation might be carried out as follows – the direct ray is attenuated to be equal to the echo, and delayed by the device so it appears at the same place on the scan as the echo. The delay will then measure the range.'

In case his readers had still not got the gist, he allowed himself some more explicit speculation. 'The writer's (personal) opinion is that the installation was a 60 centimetre RDF.' It was radar. He added, 'It seems strange that no one was curious (before January 1940) about the "aerial" on the control tower.' The lack of curiosity, however, would continue.

The report was filed and then forgotten, seen by some officials, understood by fewer, and then left in the archives of Whitehall. Britain continued for at least a year to believe that it, alone, had mastered this new wonder weapon of radar.

In his hotel room, Langsdorff had one last task to perform. This task, he explained in his letter to the German government, was a moral and military necessity. 'For a captain with a sense of honour, it goes without saying that his personal fate cannot be separated from that of his ship,' he wrote. 'I can do no more for my ship's company. Neither shall I any longer be able to take part in the present struggle of my country. It only remains to prove by my death that the men of the fighting services of the Third Reich are ready to die for the honour of the flag.'

Placing his ship's ensign carefully on the ground, he lay on top of it. There, in the hot early summer's night, he must have felt a very long way from the December chill of his home on the distant Baltic coast. Cocking his service revolver, he shot himself.

CHAPTER 12

Heligoland Bites

'No enemy bomber can reach the Ruhr. If one reaches the
Ruhr, my name is not Goering. You may call me Meyer'

HERMANN GOERING

18 December 1939

THERE ARE ALMOST NO trees on Heligoland. In midwinter on
this small island off the north coast of Germany, it is chilly,
windy and, normally, pretty bleak. There would be little reason to live
there except for one fact: its strategic value. It is the Gibraltar of
northern Germany, guarding a naval route known in English as the
Schillig Roads. To pass along the Schillig Roads, to reach the ports of
Hamburg and Bremen and the naval base of Wilhelmshaven, you
have to pass by Heligoland.

There may be almost no trees here but, by the autumn of 1939,
locals could not have failed to notice strange new metal structures
the height of trees. On 18 December the operators of one of these
awoke to an unusually beautiful day – crisp and clear. Oberleutnant
Carl Schumacher, commanding officer of one of the fighter groups
assigned to the area, had declared it 'Splendid weather for fighters.'

His adjutant had, with suitable dramatic irony, agreed. 'The Tommies are not such fools – they won't come today.'

In Mildenhall, England, 330 miles away, it was also shaping up to be a lovely day. The night before, RAF Sergeant Herbie Ruse had received a call to return to station to prepare for a practice flight. This was the early days of the phoney war, and there was a lot of practising. But when he arrived he had been told, instead, that it was a mission – to attack shipping from 10,000 feet, to his mind a tall order. So unexpected was the attack that Harry Jones, his rear gunner, only learnt the details of the mission as they were taxiing on the runway. That morning he had left his wife Mary at home, where she was preparing their Christmas pudding, and cycled to the airfield as usual. Propping his bike outside a hangar, he went inside and learnt that he was going to be flying. He didn't have time to do much more than grab his kit and get on board. 'They've found the German Navy, we're going to Wilhelmshaven to attack them,' Sergeant Ruse shouted over the intercom, above the vibrations of the Wellington bomber.*

These were innocent, pre-Fall, days for the RAF. This was a time before the messy moral compromises – before the Dresdens and Hamburgs. Like their fathers in the trenches of 1914, shaking hands and playing football with the enemy on Christmas Day, these aircrew and their masters still thought civility in aerial bombing was possible.

Days before Britain entered the war, Franklin D. Roosevelt, the US president, sent a message 'to the Governments of France, Germany, Italy, Poland and His Britannic Majesty'. It is, today, both prophetic and, given what would come to pass over the next six years, poignant. Referring to, among other tragedies, the bombing of Guernica in the Spanish Civil War, he wrote:

> The ruthless bombing from the air of civilians in unfortified centres
> of population during the course of the hostilities which have raged
> in various quarters of the earth during the past few years, which has

---

* Much of these accounts are pieced together from the original reporting in Max Hastings's masterful *Bomber Command*.

resulted in the maiming and in the death of thousands of defense-less men, women, and children, has sickened the hearts of every civilized man and woman, and has profoundly shocked the con-science of humanity.

If resort is had to this form of inhuman barbarism during the period of the tragic conflagration with which the world is now con-fronted, hundreds of thousands of innocent human beings who have no responsibility for, and who are not even remotely participating in, the hostilities which have now broken out, will lose their lives.

Roosevelt called on all sides to renounce such tactics. Britain not only pledged adherence to this principle but, a mere three years before trumpeting 'thousand bomber raids' on German cities, did so enthusiastically. In the hours before the official declaration of war, the cabinet met to thrash out bombing policy. There were, it was decided, two options.

If the Germans restricted themselves to military targets, abiding by the Roosevelt doctrine, then the RAF would do the same. It would target the German fleet at Wilhelmshaven if and only if its warships were at sea – it didn't want to risk a stray bomb killing civilians. It would also drop propaganda leaflets over cities, an action which, the cabinet's minutes attest, 'would have an important effect on German public opinion'. (Later, Arthur 'Bomber' Harris said the main effect of these leaflets had been to 'provide the enemy with five years free supply of loo paper'.)

If, however, the Germans started targeting civilians, then the RAF would re-examine its scruples. As the cabinet minutes attest, it would 'attack power supplies, oil resources and war industry in general – unhampered by the consideration that they will inevitably involve incidental loss of life to civilians'. But the idea of targeting civilians directly? Well, that was monstrous, inhuman and un-British. 'We wish . . .' the consensus statement continued, 'to record our convic-tions that if we adopt the second course we should do so in full and without any restrictions – other than those that we shall always observe, such as refraining from attack on civil population as such for the purpose of demoralization.'

On this basis, the rules of engagement for Harry Jones and the twenty-three other bombers taking off beside him were explicit: 'Task: to attack enemy warships on the Schillig Roads or Wilhelmshaven. Great care is to be taken that no bombs fall on shore.' As Harry and his colleagues reached 15,000 feet over a deep blue North Sea, bashing their gloved hands together to keep the circulation going, they knew that if there was the slightest chance of hitting a civilian they were to abort. What they did not know, because they did not think it possible, was that they were already being tracked.

At Heligoland, the radar operators spotted the bomber force approaching as soon as it was in range and called it in. According to the accepted orthodoxy in the RAF, only Britain had developed a working radar system. In fact, though, the Heligoland station was only one of several German radar systems that followed Harry Jones and his colleagues that day. So too did another, among the sand dunes of the north German coast.

The German radar operators were, initially, disbelieved. Visibility was near perfect, the weather was calm, the low winter sun provided perfect cover for approaches from Luftwaffe fighters to the south. Who would risk a massed assault in such conditions? The fighter controllers could not initially believe what they were being told. 'You're plotting seagulls!' came the incredulous reply. This dismissal meant that ninety minutes after being spotted, the bomber force reached their target, incredibly, unmolested. There, they opened their bomb-bay doors, overflew the German fleet, and didn't drop a bomb. The ships were deemed too close to the shore – all but a handful of the Wellingtons failed to find a target.

It was missions like this that the editor of *The Aeroplane* was thinking of when, in early 1940, the magazine wrote, 'Some amazing stories of the opportunities forgone by Great Britain in observance of the law will be told some day. Pilots, confronted with perfect targets, have had to keep the law, grind their teeth in chagrin, and hope for a change in the temper of the war.' There was a modicum of teeth grinding among the crews of the Wellingtons. But they had little time for chagrin. They were three hours from home, and half of them would never reach it. By now, no one in the Luftwaffe thought they were seagulls.

The Messerschmidt 109s and 110s were now as relentless in the pursuit of the retreating bombers as they had been hesitant during their advance. Thrusting and pulling back, thrusting and pulling back, they brutally exposed the inadequacies of the bombers' defences. When they made it to Herbie Ruse's Wellington, towards the rear, Harry Jones tried to shoot but his gun was jammed, frozen in the December cold. He later remembered looking across to see another Wellington, its rear gun crewed by his friend Len Stock. The entire back of the bomber had been shot out – including Stock.

There was no pretence at keeping up formation. Ruse pushed his stick forward, plunging his plane into a steep dive. But it wasn't enough. As Jones desperately tried to get his gun working, the fighter returned, raking the bomber. One bullet scraped his back, another shattered his ankle. 'Can you do anything back there?' shouted Ruse over the intercom. 'No? Then for God's sake get out of the turret.' He crawled into the fuselage, where his crewmate Fred Taylor tried to administer morphine. Even as he was preparing to plunge the needle into Jones's leg, another burst of machine-gun fire ripped through the bomber and shattered Taylor's skull.

The plane was doomed. Ruse pulled out of the dive, to see the dunes of the north German coast rushing beneath. The Wellington crashed into the beach, churning up the frozen sand. Dazed and wounded, the three survivors sat and awaited capture.

With the destruction of half the British raiders came too the destruction of the idea that this would be a force of daylight bombers. The idea that Britain, alone, had the means to use radio waves to detect enemy planes, however, somehow survived to fight another day.

This was not the first disastrous raid on Wilhelmshaven. On 4 September, the day after war was declared, a crew of fifteen bombers listened to a stirring message from the King, and then took off to carry out the first bombing sortie of the war – attacking the German fleet. Five would not return.

Then, two months later, and a month before another futile and costly raid, a document turned up at the British Embassy in Oslo – a document which only Jones took seriously. The author of the Oslo

Report 'maintained that in the raids on Wilhelmshaven in the first two months of the war, our aircraft had been detected at a range of 120km by the RDF [radar] installations which covered the NW German coast,' wrote Jones, in the first of many official reports that mark his struggle to persuade his superiors they were fighting an enemy protected by radar. To little avail. 'This statement,' he continued, 'was not generally accepted in this country owing ... in part to a widespread disbelief in the existence of German RDF.' There then followed, he later recounted, a 'subsequent long period of comfortable disregard'.

During this period, the evidence accumulated on British desks. Some reports, like that from the inspection of the *Graf Spee*, Jones considered conclusive. Much was circumstantial. There was an engineer for AEG in Berlin who was reported by intelligence sources to have stated that they 'were supplying the German Army with camouflaged trucks bearing high frequency and high tension units driven by petrol generators. It was believed by our source that the apparatus was intended to "direct electric rays".

A refugee from Germany claimed that 'the Atlas works of Bremen were experimenting with reflected radio pulses at 800 yards from a wall in 1931.' A prisoner, wrote Jones, said that British radar apparatus 'is considered somewhat antiquated because its need of high towers suggests that the German system operates on much shorter wavelengths.' Another report referenced a 'Fliegersentfernungsmessgerät', a delicious compound word that translates as 'flying thing distance measurement machine.'

Jones kept returning to the urban myths that he had heard from across Germany, of the radio masts that stopped car engines. 'The only certain known facts apart from the external nature of the towers, was that work was done there for the Luftwaffe, which was maintained secret even from the normal personnel.' In the case of one mast, based at a broadcasting station in Frankfurt, they had learnt a civilian operator 'found himself in trouble with the Gestapo for taking too much interest'. 'It is natural,' said Jones, 'to suspect RDF but no evidence has come to light of any likely transmission.'

There were also hints as to the German capabilities, if this was indeed radar. British intelligence managed to get its hands on a paper that instructed Luftwaffe bombers when they were attacking Britain to assemble over the French mainland rather than the sea, 'because the special British wireless stations would be ready in action and reporting the assembly.' Was this, thought Jones, because they knew their own radar could do the same? 'The Germans therefore at least suspect that we can detect them at the distance of the French coast: this suspicion may be based on operational observation and/or knowledge of what they could do themselves.'

Then, on 5 July 1940, an intercepted message landed on his desk. It was an account of a report, in which the Luftwaffe stated it had been able to intercept RAF reconnaissance flights 'owing to excellent Freya-meldung'. This was when, said Jones, the 'long period of comfortable disregard was ultimately broken . . . It was obvious that Freya represented something new.' But what? *Meldung* is the German word for report, and Freya is a female name. Or, more pertinently to Jones at least, the name of a pagan goddess. It could be nothing. Or it could, Jones suggested, be the vital clue. 'The example of Wotan left little doubt of the association with the Norse goddess,' he wrote in a report. He went to Foyles bookshop and bought a book on Norse mythology. What he then wrote on the basis of his research can either be viewed as a masterpiece of deduction, or of motivated reasoning.

Freya was the Norse goddess of beauty and love. To Christians, for whom the meek and virginal Mary was the model of womanhood, she was both a shocking and titillating figure: blonde, milky-skinned, stunning, but also sexually assertive. 'Few of her attributes have any possible relation with the present problem,' conceded Jones, in a supposedly technical report that had rather more literary – and sexual – allusions than Whitehall bureaucrats were generally used to. 'She did, however, have as her most prized possession a necklace, [called] Brisinga-men, to obtain which she not merely sacrificed, but massacred, her honour. The necklace is important because it was guarded by Heimdall, the watchman of the gods.'

This, Jones maintained, was the crucial clue. Heimdall's great power was in his senses. With his ears, he could hear the grass growing in the earth and the wool growing on sheep. With his eyes he could see a hundred leagues. He never slept. Here, straddling the earthly and the divine, he used his talent to warn of invaders. Behind him, the gods could feast – secure in the knowledge that they would not be caught unawares.

'There is a possible association of ideas with a coastal chain and a detecting system with a range of 100 miles.' Heimdall guarded Freya's necklace with his superhuman senses and eyes that could see a hundred leagues. Jones suspected that the Germans were guarding their own precious Reich with a device that, like Heimdall, could see to the horizon. Was this what Freya referred to? There were other hints. 'Moreover,' added Jones, 'in Germany, the Brocken is pointed out as the special abode of Freya, and the mystery of the tower on the Brocken is well known.'

It took a certain sort of mind to see a Luftwaffe intelligence report that mentions the codeword Freya and make the leap not just to an erotic goddess, but then on to her loyal watchman – and from the watchman to radar. If it hadn't proved to be true, it would have been viewed as desperate stuff. Jones conceded in the report that an over-reliance on Norse mythology to provide insights into electromagnetic warfare might be a bad idea. He also conceded that on the basis of his own reasoning the codeword 'Heimdall' would have made more sense, but then added, somewhat bafflingly, that the Germans might have rejected it, like a crossword-setter finely tuned to their audience, for being too obvious.

The truth is that Jones didn't need any of the Norse analogies. He was already convinced the Germans had radar. He had seen the description of the *Graf Spee*, he had read the Oslo Report. Just as he had done with the beams, his job was to convince others of the truth he already saw – that as well as using radio waves for guidance the Germans were, like the British, bouncing them off incoming planes as early warning.

In his view, the key information in the Freya report was not actually the name, but the other details provided alongside it. The agent who handed over the information had established that the apparatus had spotted planes, was in use in two separate locations and was protected by portable anti-aircraft artillery. 'The contribution of this source amounts to the facts (1) that the Germans have an aircraft detection system, (2) which is an aid to interception, and (3) which depends on the Freya Gerät which (4) merits protection 3.7cm Flak, and of which (5) two positions have been given, both of which are on the coast,' wrote Jones. 'It is difficult to escape the conclusion that [Freya] is a form of portable RDF.'

Even at this stage in the war, after everything they had seen the Nazi war machine achieve, the British were not prepared to concede that German scientists might have matched British ones on this technology.

In the Oslo Report they had a document that described radar. Even if the report was the misinformation many dismissed it as, the Germans clearly knew how radar worked in theory. In the *Graf Spee*, they had proof the Nazis also knew how to make it work in practice. Now there was a document describing a device that could spot British planes and direct the Luftwaffe to intercept them. A logical conclusion would be that Britain did not have a monopoly on plucky and ingenious science.

Whitehall did not reach that conclusion. Instead, a hurried series of telegrams were sent. Yes, it was possible the Germans had radar. Yes, if they did it was already operational. But they must, surely, have developed it through underhand means?

Churchill fired off a memo of his own. 'Ask the Air Ministry whether any [British] RDF stations fell intact into the hands of the enemy in France. I understand there were two or three. Can I be assured they were effectively destroyed before evacuation?'

Churchill's theory was correct in one respect. The invading German forces had indeed captured a mobile British radar set. They had found it, inspected it, and dismissed it on the grounds that it was significantly cruder than their own version.

CHAPTER 13

Proof Incontrovertible

'Some day we're going back'

R. V. JONES

February 1941

IT WAS FIFTEEN MONTHS after the Oslo Report, fourteen months after the sinking of the *Graf Spee* and six months after Jones went to Foyles to buy a book on Norse mythology to see if he could work out what Wotan and Freya got up to. Air Chief Marshal Joubert, the RAF officer in charge of radar and signals, among other things, had called a meeting and there was one item on the agenda: 'To discuss the existence of German radar.' Did the Luftwaffe, they wanted to know, really have this technology?

This time, Jones wanted there to be no doubt. In the preceding six months, and in particular the preceding six weeks, he had been busy. It had become clear to him that the Germans not only had radar, but were using it effectively.

One breakthrough had been in an academic field that was sometimes almost as complex as electromagnetic science. Namely, acronyms. In those early years radar went by many different names. There was Freya, of course. There was also DT, standing for *Dezimeter-Telegraphie*, the generic German name for radar technology, and Heavy DT. There was EEMG. There was even Englisch DT – a nod both to the fact that the RAF had used radar defensively first and that the Luftwaffe were entirely aware of this fact. The multiplicity of names served to hide the bigger picture. When Jones at last deciphered each term and realized they were descriptions of the same thing, it was possible to piece it together, linking all the different Enigma intercepts and agent reports.

Then, he could make sense of the past few weeks and months. In particular, there was one report about an operation conducted on 29 July by a 'DT station' on Cap de la Hague, near Cherbourg. That was the date when HMS *Delight*, a destroyer, had contravened orders and sailed out of Portland Harbour in daylight. When she was just twenty miles off Portland Bill, she had been spotted by some Stuka dive bombers, who attacked. There was one direct hit, and it was enough – she sank, with the loss of six crew.

Except, Jones had realized, she wasn't spotted by the Stukas at all. She was spotted by the station at Cap de la Hague. Not only was this useful intelligence, it was a worrying sign of just how advanced German radar was. 'As *Delight* had never been nearer than 60 miles to the station and had neither air escort nor balloons, this claim might appear incredible if it were not well authenticated,' he later wrote. Yet, still, he lamented, 'there remains some expert prejudice against believing the Germans had radar.'

The truth, in fact, was even worse. At around the time that Jones was linking the different acronyms, Ludwig Becker, a German night-fighter pilot, was about to make history. So too – although they would never know it – was the crew of an RAF Wellington.

The night of 16 October 1940 was bright and moonlit. Ten thousand feet up, off the north coast of Nazi-occupied Holland, Becker was flying a Dornier DO 17Z-10 nightfighter. Until now his job had been lonely, and difficult. How, amid the vast blankness of the sky, do you spot a plane, your ostensible target, travelling at 250mph? When there are clouds, when your quarry can move in three dimensions, when the towns below are blacked out, what hope of tracking down the enemy?

Some of his fellow pilots, faced with this impossible task, had fallen into despondency. Among them was Helmut Lent. A year earlier Lent had been the toast of the Luftwaffe, taking three kills in the incident that had become known as the Battle of Heligoland Bight – the ill-fated RAF assault on the fleet at Wilhelmshaven. That was during the day, though. Now tasked with protecting the Fatherland by night, he felt his skills were being wasted in a pointless pursuit. He went to his commanding officer to ask for a transfer.

'I request to be re-mustered to day-fighting, Herr Major,' he said. 'I just can't see at night.'

Herr Major convinced him to stay, to give it another month. Lent did, and would once again become the toast of the Luftwaffe – taking 102 kills by the time he died in 1944. A large part of his subsequent success comes down to what would happen on the night of 16 October with his colleague Becker.

In Becker's cockpit that night, he and his crew were no longer quite so lonely. Instead, they were in constant contact with a Freya site below. That Freya had picked up a target Becker had missed, and a radio operator was following both planes and guiding him in. It worked. 'Suddenly,' he recalled, 'I saw an aircraft in the moonlight about 100 metres above and to the left; on moving closer I made it out to be a Vickers Wellington.'

Here, at last, was his quarry. Stealthily, he positioned himself for the kill. 'Slowly I closed in from behind, and aimed a burst of 5–6 seconds duration at the fuselage and wing root. The right motor caught fire immediately, and I pulled my machine up. For a while the Englishman flew on, losing height rapidly. The fire died away but then I saw him spin towards the ground, and burst into flames on crashing.'

It was the first time in the history of warfare that radar had been used successfully in this way to guide an aerial attack. Becker could not see in the dark, so he borrowed a different eye – an eye that saw a different kind of light. He was a blind man, and this radar station was his guide dog. Or, perhaps, his attack dog.

Thus far, it was a crude system. The Freya could not judge altitude, so a lot of luck was involved. But a new companion technology was coming into service, called Würzburg, that would change that.

Würzburg did not have the range of Freya. What it did have was the ability to judge not just the distance and direction to the target, but the vertical angle too. Würzburg was a three-dimensional tool, for a three-dimensional war. And Jones knew almost nothing about it. But as the RAF went on the offensive, as it prepared to take the battle to the enemy, uncovering the enemy's defences was going to become his most important task. And absolutely the first, most basic thing he needed to do was convince his superiors those defences existed. He needed all the intelligence sources he could find – however unlikely it was they would yield useful information.

Towards the end of 1940, a new breed of agent had been recruited and dropped across occupied France. 'From the counterintelligence viewpoint,' these agents were, wrote intelligence officer Brigadier General Thomas J. Betts, 'very safe'. They 'do not possess national characteristics, are resistant to interrogation', and of course, he added, blended right into the local population. Or as he put it: 'Once in a loft [they are] indistinguishable from local birds.' These agents were pigeons, and around their ankles were a series of instructions for the local population. There was competition for what went on these instructions. Some, those briefed in advance by Jones, contained the note, 'Are there any German radio stations in your neighbourhood with aerials which rotate?'

It was, everyone conceded, something of a punt. But not one that stretched wartime Britain's resources. Aside from their other virtues as incorruptible agents, pigeons were low cost – and, if someone did attach secret information to their legs, high reward. Over the course of the war the homing pigeons of what was codenamed Operation Columba would bring in a few tasty morsels. They were of sufficient concern to the Germans that, as in every new branch of warfare, countermeasures were brought in – in this case a division of peregrine falcons under the auspices of Goering, who considered himself responsible for anything that flew.

Pigeons did not tell Jones about Freya or Würzburg. Neither did his previously most reliable source of intelligence, the transcripts from the Cockfosters Cage. As the Blitz wound down, so too did

POW intelligence. In any case, few airmen had direct contact with radar systems. There were still inferences that could be made, though. During January and February 1941, information began to trickle in about Würzburg. In particular, Jones and Frank received the news that one Freya radar and one Würzburg would be sent to Romania, and two Würzburgs to Bulgaria.

This gave Jones an idea. 'It was probable,' he surmised, 'that these represented the minimum number to give continuous coastal cover.' The coastline of Bulgaria is 150 kilometres long, and that of Romania 260km. What he had here, as every GCSE mathematics student knows, is a simultaneous equation. If the range of a Freya, say, is F kilometres, then one Freya can cover 2F kilometres of coastline – because it can look in both directions along the coast.

This means that, for Romania:

$$2F + 2W = 260$$

For Bulgaria, covered by two Würzburgs, the equation is simpler still:

$$4W = 150$$

From this, you can work out that the range of a Würzburg must be at least 37.5km (Jones rounded up to 40km), and the range of a Freya must be at least 92.5km (Jones rounded up to 100km). 'It was therefore now possible,' he said, 'to resolve two distinct types of operators, although neither had been seen nor heard.'

But it wasn't enough. As Jones conceded, what he needed in order to persuade the doubting Thomases of the Air Ministry was to see them, and hear them. A different form of intelligence, less flighty than pigeons, was required. There was no option other than to try and spot them from a plane.

If you tether a goat or a cow, simple geometry dictates that it will graze in a circle. That circle, made by and containing a blameless ruminant, will come out clearly on an aerial photograph. And, oddly

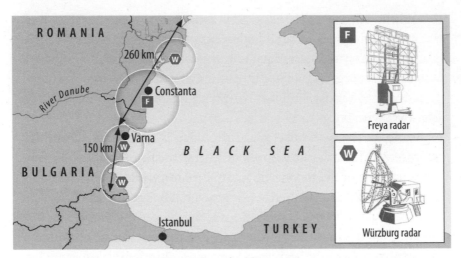

The coastlines of Bulgaria and Romania gave Jones key clues to the range of Freya and Würzburg

Würzburg radar

enough, it will look a lot like a gun emplacement. This was why, at the start of the war, Britain's fledgeling photographic interpretation unit – tasked with analysing the images taken by Spitfires zipping across the skies of occupied Europe – reported a surprising number of artillery batteries in oddly non-strategic, but always agricultural, locations.

The world, it turns out, looks very different from above. And the interpreters of this world, who would during the second half of the war become as important to Jones as the POW interrogators had in the first, had to create their own science from the ground up. Or, rather, the air down.

It was about spotting the unusual even when it looked mundane. It was about assessing shadows – in which hidden details of ships and armaments became apparent. It was about noticing when an apparently harmless location had changed subtly and significantly between shots.

Michael Spender, one of the earliest employees, who was instrumental in spotting barges preparing for cross-Channel invasion in 1940, said it was about seeing what wasn't there as much as what was. 'An interpreter is like a motorist driving through a town, who suddenly sees a rubber ball bouncing across the road from a side street. He can't see any children playing but he knows in a flash they are there and his brake is on. You must know what is normal, but you must also know the significance of what you see when you see it.'*

Douglas Kendall, who headed operations, likened it instead to puzzle solving, where the fragmentary picture held all the clues to the whole. 'In the jungle,' he said, 'it is not necessary to see the whole elephant to know that the elephant is there.'

Later in the war, guidance was produced for new interpreters in one of their hardest but most crucial tasks: spotting radio technology. Finding the enemy's antennas and transmitters was, the document explained, among the most important work. 'Interpretation of wireless apparatus brings us immediately into the front line of operational

---

* For the definitive account of RAF Medmenham read *Spies in the Sky* by Taylor Downing.

warfare. Day and night unseen but ceaseless combat is proceeding and the fortunes of the fight are revealed in the new apparatus constantly being installed,' rookie interpreters were told.

'Wellington said that the successful general can see what is happening on the other side of the hill. Photographic reconnaissance performs this service over long terms, but the actual operational contacts are directed to an increasing extent by wireless. An immense organization employing thousands and thousands of men now controls the movements of day and night fighters.'

For all the complexities of the equipment they were seeking, though, just one pertinent fact about the technology needed be known for them to perform their job: 'Namely, that the length of the aerial should be a direct function of the wavelength.' If they knew how high the antenna was, they instantly had useful information. You cannot tell height from above but, on a sunny day in the morning or evening, you can infer it from shadows. If you know the time, the date and the latitude, and you assume the ground is flat, then the length of a shadow tells you exactly how tall a radio mast is.

Of course, those calculations are fiendishly complicated. To help them along, one of the interpreters produced a mechanical computer – closer in design to a medieval astrolabe than the devices being made at the same time at Bletchley. A small globe made up of interlocking circles of wood, it became known as the Altazimeter – and so important did it become that one of the interpreters wrote a poem in its honour.

> O wondrous spheroid gadget, now complete
> With scales and whatnots properly adjusted;
> Designed to help when e'er the brain with heat
> From calculating heights is nearly busted.

Since July, Jones had been working with the interpreters seeking out evidence of radar, using their Altazimeter to look for the tell-tale shadows that indicated an installation was there. But to no avail.

Then, in the new year, two photographs came in. The winter sun had cast a thin shadow from circular objects on Cap de la Hague in

France. They were about twenty feet across, and could well have been cattle-feeding troughs. They were also right where they knew a Freya station was located. The plane providing the photographs, flying far overhead, had taken two shots nine seconds – or, at the speed the plane flew, nine-tenths of a mile – apart. This was deliberate. The idea was that the images could be used together as a 'stereoscopic pair'. With the left eye seeing one shot and the right seeing the other, the interpreters would be given a 3D view. It would be as if they were looking down from above, but as great giants with eyes just under a mile apart. This technique only works if there is no movement. If something has changed – a car has moved or a tree has swung in the wind – then rather than appearing three-dimensional to our brains, that part of the image just looks fuzzy.

In these two images, joined together into one stereoscopic whole, there was a tiny fuzzy patch, around a 2mm-long shadow. When Claude Wavell,* a brilliant mathematician and one of the interpreters, took a closer look he could see that its length didn't change but its width did. It went from 0.1mm wide to 0.2mm. There was a simple explanation for this: in the intervening nine seconds the object making the shadow had turned. Jones called the photographic reconnaissance unit and asked for a pilot to fly in closer – to perform a dangerous low pass to produce an oblique photograph.

There are immense technical difficulties in taking such a shot. The photographic pilots flew Spitfires modified to fly at the limits of what was possible. Speed was their greatest protection. But it was also their greatest challenge. To fly low enough to get a decent shot means continually concentrating to avoid crashing into bridges, hedges and pylons. It means avoiding small-arms fire from enemy soldiers close enough to take pot shots. And it means doing all this while remembering to open the shutter with split-second accuracy – because you won't get a second pass.

When the pilot returned to base, there was disappointment. He had been sent to seek out radio masts. 'He came back saying that he had seen nothing of the sort, only a Bofors gun which he had shot up,

* There is some dispute about whether it was Wavell or Charles Frank.

causing casualties. It was a gross disappointment.' More than that, for Jones it was an embarrassment. He had sent a man to risk his life – taking a photograph not only of something pointless, but of the very anti-aircraft gun that could easily have killed him. 'A more foolhardy request it would be hard to imagine,' he wrote.

What followed entered the legend of the photographic service, being cited to new recruits in the wireless detection training manual. 'However,' the account explained, 'he brought back a photo of the gun and the huts. When developed, it showed, at the extreme edge, an aerial. They had seen it.' There, in the background, was the radar they had been looking for.

Entirely coincidentally, at almost exactly the same time on the opposite side of the Channel, someone was hearing it too.

Following his successes with the beams, Jones's department had grown dramatically. Which is to say, it had gone from one to two to three. To work with him and Frank he was offered Derek Garrard, a scientist on secondment from the Telecommunications Research Establishment. Amid this vast influx of staff Jones, however, again faced his old implacable adversary: bureaucracy. Garrard's clearances were taking time to come through and he found himself with very little to do. 'But,' wrote an admiring Jones, 'Derek Garrard was not a man to remain inactive.'

Instead, he decided to head to the south coast, carrying radio listening devices. There, facing across the Channel to occupied Europe, he assembled his apparatus. He pointed a system of aerials across the sea, hoping to detect – in Churchill's famous words – the lights of perverted science beyond. And, not unreasonably, he was arrested as a spy. And while the over-zealous local police were fairly rapidly placated, it was a far from ideal outcome for a man awaiting top-secret security clearance.

When Garrard was released, instead of returning to London he persisted. Shifting his search, working up and down the electromagnetic spectrum, suddenly at 2.5m wavelength there came a signal. As he stood there shivering in the winter gloom, he heard Freya – he heard the tell-tale pattern of pulses of a radar, at the wavelength they expected. He jotted down a rough bearing, and headed home to the office. An office where, as it happened, Jones and Frank were poring over the photographs of unusual German equipment lying on just the bearing where Garrard had heard the signal.

That was the very day of the meeting called by Joubert, 'To discuss the existence of German radar.' To have a meeting of that title at all implied there was doubt. Now, though, there was none.

Jones was, at last, ready to have some fun. 'I let it run on a little to let the doubters say that they did not believe that the Germans had any radar,' he wrote, 'and then I produced both the photographs and Garrard's bearings.' Admiral Joubert was furious – thinking he had set them up. 'Jones, how long have you had that up your sleeve?' he asked. The date on the photographs, from just two days earlier, answered his question.

'That,' said Jones, 'was the end of disbelief in German radar'.

CHAPTER 14

The Flight of the Hornet

'People talk so much of what they will do. I prefer to do it'

THOMAS SNEUM

June 1941

A LIFETIME LATER, THOMAS SNEUM would remember the moment his ordeal ended – and he was finally believed. That point came when a gentle-looking giant of a man walked through the door and took over from his interrogators. This man had, Sneum would recall, a calm aura about him. 'He treated everyone as though they had some good in them, and it was his job to find it.'

Sneum needed calming, and he needed someone who wanted to find the good in him. The Dane had been shot at, by the Germans and by the British. He had abandoned his wife and newborn baby to flee his home country, quite possibly forever. He had been shunned by his old friends for being a 'Nazi-lover', and had, on a number of occasions, narrowly avoided arrest by the Nazis for activities suspiciously close to espionage. They were indeed espionage.*

* Mark Ryan's *The Hornet's Sting* covers Sneum's extraordinary life in more depth.

He had, he kept reminding his British interrogators, been so eager to bring Nazi secrets to Britain that he had flown a single-engined biplane a hundred miles beyond its range – climbing out on to the wing midway through the flight to refuel it. His interrogators, not unreasonably, didn't believe a word of it.

But now that had changed. Quietly, the man examined the film that Sneum had brought with him. As he looked through it frame by frame, helped by the occasional explanation from Sneum, the grainy form of the pictures slowly made sense. Reginald Jones looked up. 'Freya radar,' he announced. 'Hitler's latest defence system.'

It was, he said later, 'one of the most spectacular pieces of intelligence' to have come across his desk.

You could choose several places to begin the tale of Thomas Sneum. You could start it on 9 April 1940. That day, the 22-year-old pilot had been at his naval air base in Avnoe when he heard the roar of the invading German bombers above. They were at war. His adjutant opened sealed orders from King Christian and they were unambiguous: Denmark was a sovereign country and she was to be defended.

But even before he sprinted to his Hawker Nimrod those clear orders had been countermanded. Sneum might have been prepared to take on the Luftwaffe in a biplane; his superiors were not, and had thought better of it. In a radio broadcast the King had changed his mind and announced that his forces would offer no resistance. Sneum tried to push past the mechanics and take off anyway. The plane, however, had already been put out of action. Sneum was grounded. The sense of humiliation would follow him for the rest of the war.

You could start it a little later on 6 February 1941 when, after the occupation of Denmark and weeks of training during which he became adept at shooting birds out of the sky with a longbow, he stood at an attic window with his bow and waited for the car of Heinrich Himmler to pass. It never did – Hitler's SS chief had cancelled a visit at the last minute.

But as good a place as any to start is in the woods around the coast of Fanoe, an island off the North Sea coast of Denmark and Sneum's home. Here, Sneum had noticed some odd metal structures being erected. Creeping through the trees one night, he went in for a closer look. Barely daring to breathe, he listened as a plane flew over, and watched as the structures came to life. There in the dark they began to swivel, following the sound. Then, suddenly, a light lit up the sky. 'They switched on the searchlight and the beam hit the silver-coloured Junker immediately.' It couldn't have been chance. 'It was that demonstration which made me certain we were dealing with some kind of early-warning system. I was convinced that they now had the capability to plot the position of a ship or plane using radio waves.'

He resolved to return, with not just a camera but also, on the instructions of the British Embassy in Sweden, with whom he had been able to discuss the discovery, a Movikon film camera.

Sneum had been carefully cultivating friendships with local Nazi commanders, taking them to bars and getting them drunk in the hope they would reveal their secrets. In doing so, his real friendships had suffered. Now he had no need any more to spend evenings being glowered at by barmen and spat at by former friends. He had the secret he needed right here. He knew, though, that the next time he visited the radar installation he needed to return in the day.

Sneum could only guess at the site's importance. Across the North Sea, far beyond even the range of the swivelling radar of Fanoe, Jones and his colleagues had no doubt. There, from their offices in the Air Ministry, they were trying to piece together the jigsaw of German radar defences in the hope of mitigating Bomber Command's appalling losses. Unlike Sneum, all they had were half glances and educated guesses.

No longer did they doubt that a sophisticated defence system had been erected – tales were coming in from across the continent. But how to interpret them? 'Many agents were searching for stations.' They were seeing them, from afar, or in peripheral vision as they pretended not to look. But few agents were able to send coherent descriptions of the apparatus. It is hard to describe in words something you don't understand.

'Their appearance stretched the descriptive powers of the agents of the low countries. "Upside down umbrella" "inverted mirror" "magic mirror".' One site, which was later identified by a reconnaissance flight as a radionavigation beam device, was variously described in reports as an underground aerodrome, a 300m-high wireless mast and a long-range gun.

The bravery of these French, Belgian and Dutch agents whose reports were fed back to Jones – many of whom would die – was not in doubt. But their technical knowledge was. Jones needed to see these sites for himself, close up. Or, as a second-best option, he needed a film of them in action. He needed a Sneum.

Sneum cycled down to the trees and dunes to the north, accompanied by a resistance sympathizer called Peter; whose surname, like so many of the details of the men and women who showed such heroism, is lost to history. Both were carrying shotguns. Not for defence – you don't shoot your way out of Nazi-occupied Europe – but as cover. If caught, their plan was to say they were shooting rabbits.

Before he could even get into position among the tree cover, Sneum saw the installation begin to creak and swivel – following an unseen and distant plane. He took a risk and rushed off his bike to begin to film – the reel whirring in the Movikon.

Suddenly, behind him came a hissed warning from Peter. Someone was coming. Tucking the Movikon away, Sneum turned round to see a German guard, rifle raised, approaching. He had to think fast.

'I crouched down in the tall grass. In that squatting position, my knees were sticking out.' That covered the bulge from the Movikon. 'At the same time I pulled down my trousers and pants.'

Still pointing his weapon, the guard 'got close and shouted, "What are you doing here?" So I replied, "I'm taking a shit." He looked embarrassed and said: "Oh, OK".'

Sneum prudently decided to leave. He would have to visit another time to get the shot he wanted – his second trip turning out to be only marginally less hairy. Then, he just had to find a way to get the images out.

The original plan had been to send the film to the British Embassy in Sweden via a courier. He now decided the films were too bulky to risk it. At least, that was his excuse. It seems reasonable to question whether a sober risk-benefit analysis, as opposed to a taste for adventure, was ever his main motivation.

He resolved to return to the skies. He had intelligence that was, he was convinced, crucial to the war effort. He also had a plan for getting it out. It involved an old barn, a wreck of a plane, and an audacious escape that – in a conflict not short of audacious escapes – stands out for its audacity.

After his successful return from the secret installations of Fanoe, Sneum had begun hatching this second scheme. He had travelled to the mainland and visited a Copenhagen dealership for de Havilland planes and stolen their customer list. Then he had approached the most likely prospect – a farmer and army reservist with an ageing Hornet Moth – and explained his plan. The farmer had offered it to Sneum for free, but with the caveat, 'If you're caught, I'll say you're a thief and claim I've never met you.'

In the farmer's barn, he at last had his plane – and a friend, Kjeld Pedersen, whom he had recruited as co-pilot. True, the plane had not flown in a long time, and when the farmer showed it to them, Sneum's excitement was slightly diminished by the discovery its wings were not attached. But these were minor inconveniences. Sneum had, at last, a part to play in the war. He had written a last letter to his wife and daughter. He was not a family man and their short marriage, which only began because her furious father discovered the pair were sleeping together, was already all but over.

There in the field, Pedersen and Sneum worked into the night to get the plane ready, improvising bolts that would (they sincerely hoped) keep its wings attached. Then they waited. In the distance, quiet at first but growing louder, there came the sound of an approaching train. It wasn't much, but it was all they had to mask the noise of the engine. They spun the propeller, and heard a throaty burr as it caught.

That midsummer night, after years lying unloved on a Danish farm, the Hornet was going to fly. Sort of. Bumping over the turnip field, taxiing to the strip they had designated as a runway, the plane kicked up dust – with Pedersen running beside to help watch out for obstacles. If Nazis came, he had his revolver cocked and 'a crazy look on his face', Sneum said. 'He had told me that he would kill as many Germans as he could with his pistol and the rest with his bare hands.'

As they swung into the field, facing down a slight incline, Pedersen jumped into the Hornet and Sneum opened up the throttle. The plane accelerated, roaring down the hill – and failed to take off. Twice it would float, tantalizingly, above the grass, then return with a bump. It was too heavy, they had too much fuel. All the time, they were eating up the distance between them and the electricity pylons at the end of the 'runway'.

When, achingly slowly, the wheels at last lifted, they realized they did not have time to clear the wires. Keeping the nose down and the engine revving, Sneum instead flew under them. Immediately after, the cables zipping above their heads, he faced the next obstacle: a railway embankment with, he saw, a train on it. The driver and firemen were so close he could look into their eyes. 'They were looking as though we had just fallen down from the moon', he said. He pulled up as hard as he dared and they zipped past the cabin, with metres to spare. The immediate danger had passed. The greater danger was yet to come.

It is worth reiterating at this stage just how foolish the escape was. Sneum had decided to fly to Britain in a slow and ageing plane because he had come across evidence that the Germans had sophisticated installations – for spotting and then targeting fast and modern planes.

When the pair passed over the radar installation that they were about to betray, this point was rather forcefully brought home to them as flak shells exploded around them. The fact that a fighter plane did not follow was – they thought as they rushed to gain height – a miracle. In fact, they would later learn, it was because far to the east Operation Barbarossa was about to begin, and every available plane had been sent away for the invasion of Russia.

Of all the people in the world, few had a better understanding of how lucky Sneum had been at that point than the small team several hundred miles away in the warmth of the MI6 HQ.

Ever since the existence of German radar had, finally, been accepted, Jones, Frank and Garrard had been watching the erection of a great defensive line on the other side of the Channel, its scale and ambition easily matching that of Chain Home in Britain.

In this task they had been greatly assisted by a bafflingly basic security lapse. When sending secret information, it is greatly preferable to use landlines, which can't be overheard, rather than radio, which can be intercepted by anyone within sight of the tower and perhaps well beyond. This was why each of the Freya stations, they learnt, had had landlines installed. In one case, they had tracked the laying of fifteen separate connections.

Yet as Jones, Frank and Garrard discovered, more often than not the operators still used radio. For reasons they could not quite understand, 'security,' wrote Jones, 'was sacrificed to simplicity,' and when the Germans spotted a plane they broadcast its position over the airwaves.

When an RAF photographic reconnaissance plane flew over occupied Europe, it took photographs that placed it at a set position at a set time. When a Freya station spotted it, it sent a message to headquarters, in very basic code. The Germans thought that with these messages they were fixing the position of the RAF planes – which could then be passed on to fighter squadrons to intercept them. And, indeed, that was what they were doing. But since the RAF knew where their planes were and were listening in, when the Germans sent the messages indicating the bearing on which each station had spotted the planes, to the listening British they were in effect unwittingly revealing the positions of their own stations.

'It has therefore been possible,' said Jones, 'to watch the main chain in operation and ask questions of the enemy by sending out test flights and waiting for him to report them.'

From these direct observations and other intelligence, they had identified around 50 Freyas, in a chain from Kirkenes, on the northernmost tip of Norway, to Bordeaux, and then around the Mediterranean to the Balkans. Jones estimated this was about a third of the true number. From sources inside Europe they had learnt the serial numbers of four radars: 22, 59, 82, 132. 'We can only assume that these were selected at random from the total number of Freyas in existence, and working backwards, [the total] should be of the order of 150,' wrote Jones.

This work was crucial. As Britain found time to breathe, as the nightly Blitz wound down, understanding the Nazi radar defences had become a priority. But precisely because the Blitz had ended, Jones's job was getting harder. The Cockfosters Cage, where downed airmen were brought to be interrogated, found it had vacancies.

Jones liked Luftwaffe prisoners, regarding them as 'generally more intelligent and inquisitive than those from the other services'. Now, though, supply was drying up. 'In the purely air offensive of Bomber Command these convenient sources were almost entirely denied, and intelligence had itself to become offensively minded and go out to get its raw material.'

It was imperative that Britain found other ways to see into the mind of the enemy, and in particular into its air defences. The country had no continental army that could match Germany, and it could do little more to defeat it with its navy – which meant, Churchill had declared, there was only one choice left for retaliation.

'There is one thing that will bring him back and bring him down, and that is an absolutely devastating, exterminating attack by very heavy bombers from this country upon the Nazi homeland.' To do that, the bombers needed to know what they were facing.

Slowly, exactingly, the photographic interpreters had honed their skills at spotting the radar. Looking for rotation had proved crucial – and was possible by comparing images taken closely together. 'Shut alternate eyes and see if movement is noticeable between successive exposures,' advised a guide for new interpreters. Although, it cautioned, 'Movement, it must be remembered, is not conclusive proof of RDF as it occurs also in the case of the sewage disposal tanks.'

There is, though, only so much information you can glean from above. To really understand a system like Freya, you needed to see it up close. You needed Sneum's footage – footage that might have escaped the attention of the Luftwaffe's fighters, but was still far from home and dry.

The next occasion on which Sneum and Pedersen should have died came when the engine gave out. 'Our oil pressure indicator suddenly dropped to zero,' Sneum later recalled. 'Thinking it would only be a matter of minutes before the engine seized up, we wished each other goodbye and agreed that the effort had been well worth trying.'

Gliding down to the North Sea, they prepared for impact. 'We were both positive it spelled the end . . . it is one thing to live together as pals, and another to die together.'

But then, like a miracle, the engine puffed and blew, and began to hum. Once again, it felt like a divine intervention. Later, just as Sneum had discovered the lack of enemy fighters was due to the invasion of Russia, this miracle would also be explained to him – at altitude, it was likely that ice had formed in the carburettor. When they lost height, it melted.

He had no time to consider the thermodynamic niceties of their salvation. Now, just 150 metres above the chilly, frothing North Sea, he needed to carry out a yet more dangerous manoeuvre. He needed to refuel the plane, using the drums they had carried with them. Handing control to Pedersen, he forced open the door, into the rushing and howling wind.

'I stepped out on to the wing with my right foot and held on to the inside of the door frame with my left hand.' Carefully, the struts juddering in the wind, he passed a tube out of the cabin and, with freezing hands, undid the fuel cap and thrust the hose in. Then he flung himself back into his seat, and began funnelling in petrol – the fumes filling up the small cabin and making them retch.

Finally, as dawn broke, they spotted the coast of Northumberland. And the coast of Northumberland spotted them. Once again, with not a little irony, radar threatened to be their undoing. Soon the pair found themselves surrounded by RAF planes, indicating that they should land.

Below, the local Home Guard could not quite believe their luck. Presumably not considering four Spitfires and Hurricanes enough to keep a moribund biplane in check, they rushed out with rifles to take pot shots. Sneum dived lower still, hopping over hedges to avoid them. Eventually he saw an empty field and took his chance.

They emerged from the cockpit triumphant, and just had enough time to put on fresh clothes ('We wanted to be presentable, so that we would be treated like gentlemen'), before being captured – as suspected spies.

*

After the reality of radar was accepted, Air Marshal Medhurst, Assistant Chief of Air Staff, had summoned Jones. Britain was still fighting alone, and the only weapon it had to conduct that fight at scale was the bomber. 'We shall need to know everything we can about the German night defences,' he said to Jones. 'I think that we need a scientific mind to study them. I know what you were able to do about the beams – would you be prepared to take the whole responsibility for finding out how the German defences work?' It was a massive ask of anyone, let alone someone as young as Jones. Jones, never one to lack faith in his own abilities, appeared to believe himself a most prudent appointment.

When he learnt of Sneum's capture, Jones and Frank rushed down to Wandsworth, where the pair were being held in the MI5 Royal Victoria Patriotic Building. The story seemed so fantastical that no one had believed it. The consensus was that Sneum and Pedersen were clearly German infiltrators.

'There was an inevitable irony about such episodes,' Jones said. It was too good to be true that two Danes would have risked their lives in that way, and arrived with such valuable information. It was easier to believe they were spies. 'The more gallant and therefore improbable they were, the harder it was to believe they really happened.'

Frank and Jones did believe, though. A few days after the pair had been captured, they persuaded MI5 to let them go, 'and did our best to make up for [their] wretched treatment'. It was too little too late. Sneum was furious. Not only had he been imprisoned, but through an astonishing cock-up the film he had gone to so much trouble to smuggle in had been sent to be processed at the post office, and most of it had been destroyed.

'I went mad when I realized what happened,' Sneum said. '"You stupid bastards," I told them. "Do you know how many times I risked my life for those films?" They were trying to calm me but I just kept going. "Are you intelligence officers? What is intelligent about you? Why did you send them to a fucking post office in the first place?"'

Just a few frames survived. Initially, Jones could not make them out. Sneum stepped in to explain. 'I wasn't an expert but he made me feel like I was. Then we noticed a few clearly definable images. It was an exciting moment, and the scientists eagerly went to work.'

Tracing their outline, it became clear to Jones he was looking at Freya. Here at last was a picture of his radar adversary. Slowly, the puzzle was coming together. Or, at least, one half of it was.

Throughout the autumn of 1941, Jones and his colleagues were still hunting the elusive Würzburg, the more discerning companion radar to Freya. They had an idea of what it looked like; the Oslo Report had talked of 'paraboloid reflectors', broadcasting at a wavelength of 50cm. They had yet to see it. They had yet to hear it. But it is surprising how much you can work out from a little information, if you understand the physics.

Würzburg, according to the information they had from the Oslo Report, transmitted on a 50cm wavelength, a fifth that of Freya. In this way it could see finer detail, allowing the operators to plot the height of the aircraft – useful in guiding planes and, potentially, flak and searchlights. But it was also stopped more easily by the atmosphere – it had a shorter range.

The Freya, Jones surmised, would spot the approaching aircraft, and then, when it got close enough, the Würzburg would take over, locking on to provide a more accurate plot. From his back-of-the-envelope calculation, looking at the length of the Bulgarian and Romanian coasts, this handover would occur when planes were within about 40 kilometres of the Würzburg, its maximum range. This was crucial information – in more ways than one.

Radar works by sending out a signal, and then seeing if it bounces back. The time taken for the faint signal to return tells you how far away it was reflected. This means, though, that you have to transmit in pulses. If you sent a continuous signal, you would have no idea how long it had taken to travel – because you would not know which wave was the one that had done the bouncing. So, instead, you send it in batches. You fire off one pulse, leave enough time for it to return, then send off another.

This was where the range calculation came in. It told Jones what the pulse repetition rate should be – and, hence, the signal they should be listening out for. For the radar to work, each of those pulses had to be able to get back to the receiver before the next was sent. For a Würzburg with a 40km range, it has to travel 80km. A radio wave travels at the speed of light, around 300,000km a second. That means it takes one 3,750th of a second to cover 80km. So that was the pulse repetition rate.

Jones passed on the message: Britain should be listening out for transmissions broadcast on around 50cm wavelength, pulsing 3,750 times a second. Setting up a receiver pointing across the Channel, that is exactly what they found. Now, the hunt was on. Where were these paraboloid dishes, and how many were there?

The problem was, the smaller the wavelength, the smaller the device needed to be. Freya was hard enough to see. How do you spot Würzburg? The answer, when it came, seemed obvious in retrospect. You spot it because it is next to the Freya. It wasn't essential to site the two together but, given they worked in pairs, it certainly made a lot of sense.

Before Christmas, Claude Wavell, the photographic interpreter who had first spotted the rotating shadow of Freya, brought another anomaly to the attention of Frank and Jones. At the Freya site on Cap d'Antifer, on a coastal outcrop near the village of Bruneval in northern France, there were two Freyas and they were joined by a path, which also led to a large villa. Between, the path went through a field, where, curiously, it formed a loop. By that loop was a small dot.

What was it? Jones pored over the photograph, desperately wanting to enlarge it – to examine it and uncover its secrets. But it was just too small, 'so small that several photographs had to be examined to distinguish it from chance specks of dust'. He needed a better picture.

In the preceding months Jones had become friendly with Geoffrey Tuttle, the head of the Photographic Reconaissance Unit. Tuttle liked to know why it was he was risking his pilots to photograph aerials in fields. As a consequence of the friendship, though, Jones had also become hesitant about requesting flights, about ordering a young man to fly low and fast over occupied Europe in a Spitfire. 'It was all too easy to ask for photographs without thinking of the risks that had to be run to get them,' he said.

It turned out he didn't need to. Wavell had been visiting the pilots of the PRU and happened to mention the photograph. 'When you pilots follow down the French coast why don't you take a short cut across the Cap d'Antifer?' he had said. 'There is something there that interests me.'

One of the pilots was Tony Hill, an impossibly dashing man who Jones described as 'my idea of every schoolboy's hero'. A report for the photographic interpretation unit, possibly written by Wavell himself, verged even closer to hero worship, describing him as 'the most daring and talented taker of obliques that PRU ever knew'.

An oblique, a photograph from the side, was exactly what was needed. And Hill considered himself the man to do it even if, the same report noted, it 'might have been a wild goose chase and a dangerous one at that'.

On the first flight, Hill passed fast overhead, spotting something he described as like an 'electric bowl fire'. Alas, though, his camera failed. Undeterred by the fact he had now alerted them to his interest – Hill was not a man for being deterred – he just decided to have another go.

The next day, however, just as Hill was preparing to take off on an unofficial reconnaissance, his flight was halted. Three pilots from another squadron were scheduled to be taking photographs over the same stretch of coast. 'He thereupon taxied his aircraft over to the others and told them that if he found any of them within 20 miles of the target he would shoot them down,' Jones recalled.

This time, he produced an image that became a classic: a beautiful and unambiguous picture of what, to modern eyes, looks like a satellite dish. It was a Würzburg.

It was also a Würzburg, Jones couldn't help noticing, within a few hundred metres of the English Channel. He then began to look instead at the land around it – at the field in which it lay, at the cliff face on which it perched. At, more pertinently, the slope down to the sea. 'It was noticed,' he wrote, 'that although the apparatus was on top of a cliff 400ft high, there was a most convenient decline to the beach.' It was the sort of slope down which a soldier could run, or even several soldiers dragging radio equipment.

The Bruneval Würzburg installation

Suddenly, he had an idea. It offended his intellect, his view that scientific secrets were there to be deduced rather than stolen. Nevertheless, the sight of that field with its easy access to the sea meant he could not dislodge the thought. 'Look, Charles,' he exclaimed to Frank. 'We could get in there!'

CHAPTER 15

The Bruneval Raid

'Utrinque Paratus – Ready for Anything'

MOTTO OF THE PARACHUTE REGIMENT

27 February 1942

T HERE WERE MANY THINGS Major John Frost should have been thinking about as he embarked on Operation Biting – a mission today remembered as the birth of the Parachute Regiment. But it was getting increasingly difficult to ignore his bladder. Around him, as his plane approached the French coast, all was jolly among his fellow paratroopers. In fact, spirits were 'terrific', the 30-year-old later recalled. He also knew, though, that so far not a single one of their rehearsals for this mission had been successful. And, he really needed a wee.

Ninety minutes earlier, Frost had taken off from Thruxton Airfield in Hampshire in a Whitley bomber, one of twelve bound for that 'speck of dust' that Jones and Frank had first seen towards the end of 1941. It was cold, but they had sleeping bags, silk gloves and – perhaps even more importantly for morale, a generous flask of tea laced with an even more generous helping of rum. 'In our aircraft we were cheerful; we sang songs and played cards,' said Frost.

Flight Lieutenant Charles Cox, a young radar mechanic with a newborn baby and, until a few weeks earlier, the reasonable expect-ation of lasting the war without being parachuted on a daring mission behind enemy lines, joined in the singing – he performed 'The Rose of Tralee'. When the rum was passed to him, though, he was a little less eager, and had just three sips the whole flight.

In the days and weeks since Jones and Frank had spotted not only a Würzburg, but a Würzburg near a nicely sloping beach, a lot had happened. In the Air Ministry they now had strong suspicions this radar technology could be used to control flak and searchlights as well as pass on the location of bombers to the fighters. This anxiety was compounded by the discovery there were larger 'Giant Würz-burgs'. The Würzburg devices were, air chiefs concluded, 'a serious menace to our aircraft' and they needed a response. The design of countermeasures, though, was 'hampered by the lack of information about the enemy apparatus'. They needed to get their hands on one. This was Frost's job.

Frost was conventional officer material. He liked fox hunting and shooting, he was educated at Wellington and Sandhurst, his father had been a brigadier in the Indian Army. His job was anything but conventional, though. Tonight, he was to lead a group of men tasked with parachuting into France, finding the Würzburg and stealing it.

A lot had to go right. This was a kind of operation that had never been tried before. It involved the Army, the Navy and the Air Force working together. It involved precision timing, daring and imagin-ation. Military planners anticipated that even if successful, many of those on the planes that night would not be coming back. Despite weeks of intense training, each time they had practised the mission, something had gone wrong. Now there were no more practices left.

Over Le Havre, they were spotted by anti-aircraft gunners, who began targeting the small flotilla of Whitleys with flak. 'We could see the tracers coming up towards us, and it looked – being mostly orange with a little red – like a pleasant firework show,' Frost recalled. He also recalled, though, being unable to either fear it or appreciate it. 'The mugs of hot tea . . . began to scream to be let out.' Each jolt of the plane shook his bladder. Each shudder threatened embarrass-ment for the mission commander.

Somewhere in the blackness ahead, Lieutenant Euan Charteris and the nine other men on his own Whitley had stopped their singing to watch the same flak. It felt, to them, a little closer – as the explosions buffeted the plane. 'It sounded as though a man was hammering a piece of tin below us,' he said. During the flight Charteris had discovered that the man next to him had a birthday close to his. They would both be twenty-one in a few days' time, and they resolved to celebrate together. Like many of the 120 men, Charteris was Scottish. Before joining the Parachute Regiment, he had been in the King's Own Scottish Borderers. The singing continued, with, appropriately enough, a rendition of the traditional Scottish song 'Annie Laurie'.

When they approached the drop zone, though, each man retreated at last into his own thoughts. 'As you sit near the hole, it is a funny feeling; you watch the light, waiting for it to change from red to green. I felt as though I was acting in a play,' said Charteris. In the moonlight, he could see hedgerows and houses. Then the light went green, he dropped – and realized that the landmarks below him, so perfectly lit, were not where they should be.

A little under three miles away, Cox was landing exactly where he should, but feeling little better. As his parachute deflated and fluttered noiselessly to the snowy ground, he looked around and could – in that moment – see no one. Of all the paratroopers landing in France that night, Cox was the least prepared. In the public imagination, helped by an eager press, this new breed of soldier was so tough they (at least according to one Wing Commander) 'drank blood and crunched glass'. Cox was, though, here not for his glass-crunching abilities but his number-crunching ones – or, at the very least, his broader technical abilities. He was also here because, somewhat disconcertingly for him, he was expendable.

To Jones's great annoyance, when the mission had been conceived, his own subordinate Garrard had offered to accompany the commandos – arguing they could benefit from a technical specialist. This was annoying because, said Jones, 'as soon as he had volunteered, I had also to do so'. This was not a prospect he found appealing. Luckily, Lord Mountbatten, in charge of the operation, declined both scientists. The military wasn't so incompetent that it would send untrained boffins in possession of the country's greatest secrets behind enemy lines. Mountbatten did, though, like the idea of having a technical expert on board.

So Charles Cox put himself forward. Sort of. Cox had been summoned to a meeting with Air Commodore Victor Tait at the Air Ministry. 'You've volunteered for a dangerous job, Sergeant Cox,' said the commodore.

'No, sir,' said Cox.

'What do you mean, no sir?'

'I never volunteered for anything, sir.'

The commodore acknowledged there may have been some miscommunication, then asked Cox if he would volunteer now? Cox, not unreasonably, asked for details about the mission that was so dangerous that on hearing he had 'volunteered' a commodore felt moved to meet him in person.

'I'm not at liberty to tell you,' said the commodore, before adding, winningly, 'I honestly think the job offers a reasonable chance of survival.'

Now here he was, just after midnight in a snowy field in France, wondering just what 'reasonable' meant. 'The first thing that struck me on standing up was how quiet everything was and how lonely I felt,' he said.

He wasn't alone, though. Just a few feet away, Frost had also landed – along with some of the other big tea drinkers. They had no time for introspection. Normally, the point a paratroop force hits the ground is the point of greatest tension – a feared, dreaded moment on which much of the mission hangs. Somehow, they had collectively got themselves into the bathetic situation where instead it offered blessed relief. Finding themselves, mercifully, not under fire, the priority of the paratroopers was to christen the landing zone. 'Our first action on landing,' Frost felt moved to write in his official account, 'was to urinate.'

At just after midnight, in this corner of a foreign field, patriotic British urine splashed the benighted snow of Nazi Europe yellow. It was, said Frost, 'a small initial gesture of defiance'.

*

Before they had taken off, Charles Pickard, the RAF officer leading the flight, had pulled Frost aside. 'I expected some joke or platitude,' Frost recalled. This was when the bluff, stiff-upper-lipped pilot should have told him to give Jerry a darn good thrashing, or some such.

Instead, as they nervously awaited the signal to go, Pickard made a confession. He could not shake the worry that he was taking 120 men across the Channel to die. Explaining this to Frost, he added, 'I feel like a bloody murderer.' Frost was, understandably, unsettled.

The raid on Bruneval was about so much more than capturing a new radar – it was about proving an entirely new method of warfare. It was just the kind of operation Churchill had been itching to carry out, to at last take the fight to the enemy in Western Europe. But it was also just the kind of operation that many had warned Churchill was extremely risky.

The plan had many moving parts, but was also as simple as possible. A group including those dropped alongside Frost would take the villa at the radar site, eliminate resistance around the radar and then – ideally – remove a Würzburg radar to take back to Britain. Sappers had been trained especially in the task, given lectures on how to recognize the electrical equipment and how to avoid electrocuting themselves, and then given practical lessons in, in the words of a post-mission report, how to use 'burglars' tools'. The latter at least was a task 'at which they showed a marked aptitude'. While they were crowbarring the non-electrocuting bits of the radar out, the plan was for a second group to be securing the beach – ensuring they could get the Würzburg to the sea where, hopefully, the Navy would meet them.

What Frost did not realize, as he buttoned his flies in that field, was that they had been spotted. German reinforcements were already on their way. Neither did he realize yet that Charteris and his men – constituting half the force of soldiers who were meant to be securing his escape route – were miles from where they should be. Not that he could have done much differently if he had: there was now no choice but to get moving and hope. Frost headed for the villa, expecting a fight. 'When I got to the door it was open. I blew a simple long blast on my whistle, and burst in with six of my men at my heels, while another party simultaneously burst in through the windows on the other side of the house,' he wrote. There was no one on the bottom floor, so they ran upstairs, 'shouting "Surrender" in English and then "Hande hoch" in German. We also,' he added, rather coyly – and perhaps conscious this debriefing would be read by Churchill himself – 'used very bad language.'

At the same time, in the field outside his men were attacking the radar station, and he could hear the noise of the grenades thrown into the dugouts. So too, it transpired, could the one occupant of the villa. Completely missing the fact that he was also under attack, and presumably not hearing the 'very bad language', a German stood by the first-floor window, looking out. He was still looking out when he was mown down by a Sten gun.

The raid was daring, but the troops opposing them were not, they soon realized, the *crème de la crème* of the Wehrmacht. In a subsequent report, the British concluded they had essentially happened upon the German version of Dad's Army. 'The German troops were either very young or fairly old (say, 40–45). They were thoroughly surprised and offered little resistance. Their shooting was not good (it was mostly very high) and their sentries were not all alert.'

Even so, as Frost left the villa and ran over to the Würzburg, a machine-gun started up from north of their position – the Germans were regrouping. One of his men was fatally hit. At the Würzburg itself he found five more bodies, all German, and, around them, a heated argument. Lieutenant Young had led the attack, but now his sergeant was remonstrating with him – calling him a 'cruel bastard'.

Frost does not explicitly say what had happened. He does say, though, that their orders had been clear: 'to take only the experts prisoner and to kill the remainder'. Later, he would describe a situation where they did take a prisoner, but only on the grounds he 'had made such energetic efforts to surrender that his surrender was accepted'.

Just one of the Germans at the Würzburg had escaped the immediate attack. Flieger Heller had fled in the direction of the cliff, scrambling down in his desperation to survive. There they found him, ten feet down, clinging to the side. In the definitive early account of the attack, a book from the 1970s called *The Bruneval Raid*, Heller is a figure of fun – a bit of a dunce who had 'scuttled towards the cliff edge' in a farcical attempt to escape. In a television documentary from 1976, one soldier present described how they had asked him how many soldiers were nearby and, when they thought he was lying, had 'shook him by the lapels', after which he changed his answer.

The official files, then secret, suggest a rather more brutal, rather less *Allo, Allo*, experience. After being captured Heller, who had just witnessed his five comrades killed, perhaps in battle, perhaps in slightly more ambiguous circumstances, was indeed interrogated. Quite probably, he was shaken by the lapels. That, though, was the least of it.

How many troops were nearby, the British commandos asked. A thousand, Heller answered. It was a lie, and the British knew it. The French Resistance had, at great personal risk, passed on details of troop concentrations that contradicted this. They had established there were fifty or so troops based in the immediate vicinity of the radar site, and that a reconnaissance battalion with armoured cars could be there in an hour. They had even got the names of some officers – by copying the visitors' book at a local black-market restaurant.

So they asked again. Frost describes how one of the commandos 'hit the Hun a hell of a belt on the jaw'. Heller changed his answer to a more acceptable one hundred. Heller's interrogation was not over – but by all accounts his resistance was. Cox arrived along with the sappers and he felt the Würzburg. 'Hey!' he called. 'This thing's still hot! Ask that Jerry if he was tracking our aircraft as we came in.' Heller confirmed he was. By now the occasional shot was pinging in from the perimeter: time was running out. With screwdrivers and crowbars, they got to work.

Part of Cox's pre-departure briefing had come from Jones himself. Jones recalled with admiration Cox's bravery in the face of a very real risk that it would all go wrong. 'In my last briefing to him I warned him of the danger of his being specially interrogated if taken prisoner, and above all to be careful of any German officer who was unexpectedly kind to him. He stood to attention, smiled and said: "I can stand a lot of kindness, sir." '

Now it was Cox's chance to prove that it was not mere bravado. For three days a month earlier he had shivered on Salisbury Plain, training on a model of the Würzburg that had been mocked up by Basil Schonland, a radio scientist and, later, scientific advisor to Field Marshal Montgomery during the invasion of Europe. Now, working in the moonlight reflected off the snow, he began to dismantle the device. 'Tearing aside the rubber curtain that protected the units, we saw our prize,' he said. 'Zing! Zing! went two bullets by my ear.'

The equipment, he later noted, was 'like a searchlight on a rotatable platform'. Approvingly, he described the design as 'very clean, and straightforward'. He and a sapper got out their tools to remove it – methodically separating each component, starting with the transmitter-receiver system. He had been promised he would have thirty minutes for this task. But the intensity of fire was increasing. 'Bullets were flying much too close to be pleasant,' he said – pinging off the device. Whenever he switched on a flashlight to see a fiddly bit, 'we had to endure the protests of the parachute troops'.

They needed to get moving. Cox dispensed with a long and fiddly screwdriver, and brought out a crowbar instead. The Würzburg was levered out, the screws popping and the metal shearing. Then it was hauled on to a trolley and they withdrew. But to what? The plan had been to take it to the beach, where Charteris and his men would have secured their escape route. But half the force tasked with taking the beach had been dropped in entirely the wrong valley.

At around the time that Frost was desperately crossing his legs, preparing to unbutton his flies in occupied Europe, a few miles away Charteris was trying to work out what had gone wrong. The pilot's mistake, he realized, had been an easy one to make. For weeks before, Charteris had familiarized himself with the terrain, looking at scale models – carefully constructed from aerial photographs and from information gained from the Resistance. The valley they had landed in was surprisingly similar. But, standing in it, he realized that the trees were in the wrong place. Neither was it deep enough.

His guess was that evading the flak caused the pilot to go slightly off course. But to where? Above his head, he saw some Whitley bombers, the rest of the force they had set off with, flying on. 'I knew then,' he said, 'that we were on the right line.' If the planes were above him, then the line of their travel pointed to the drop zone. The question was, were the planes on their way to or from their destination? Had they been dropped too soon? Or too late? He took a risk, and assumed the planes flying above were yet to reach their destination. Knowing that if he was wrong it would be terrible for morale, not to mention the mission, he gathered his men and set off at a brisk pace in the direction in which the planes above had been flying. When the lighthouse at Cap d'Antifer came into view, he knew the gamble had paid off. He also knew that they were still more than two miles from their destination.

He and his nineteen men continued at a brisk march – or, as he called it, a 'fast lollop' – through the coastal countryside. But then, still some way from their destination, he realized there were twenty. In one of the more bizarre events of the night a German soldier had seen their group and joined in. Whether he was drunk, extra keen, or whether he guiltily thought he was meant to have been involved in an exercise he had somehow missed, for some distance he kept pace. Whatever slow dawning happened in the German's mind, he eventually noticed. That was when, rather than slinking off, he made his second mistake: he decided to do his duty. 'I heard a screech or a scream,' Charteris recalled. The 'Hun', he said, fired once and missed. His third mistake. He did not have the opportunity to make a fourth. Behind him, one of Charteris's men emptied 'the whole of the Sten magazine' into him.

During the march, his men had spread out, and stragglers had become lost. Charteris wrote he was unconcerned, 'They were stout fellows and marched to the sound of the guns'. Indeed they did, although, he later learnt, a few also took a diversion into Bruneval village and managed to kill three Germans with their knives. They did not have time for a diversion – they were desperately needed at the beach.

*

When Frost reached the top of the ravine that led to the beach, he heard a call. 'A voice from the beach shouted, "The boats are here. It's all right. Come on down." Almost immediately afterwards a machine-gun opened fire on us.' Sergeant-Major Strachan, five yards behind him, took seven bullets – three in the abdomen. Cox arrived soon after, with the trolley containing the Würzburg, and lay down to wait. Either his colleagues cleared the beach and he made it down, or they didn't – and he had to avoid the kindnesses of German interrogators. That was when he heard 'Scottish war cries punctuated by the explosion of grenades'. Bloodied and tired, with all the timing of a Hollywood cavalry, Charteris and his men had arrived.

They knew, now, exactly where they were. Throughout the days before they had examined this gully, learning its contours, its obstacles, its weaknesses. They had traced their route, imagined their hiding places, planned their assaults. It was utterly familiar, yet also very different. 'I felt,' said Charteris, 'as naked as a baby.' He was seventy yards from a beach house, a key strongpoint whose assault they had planned and replanned over the course of the preceding weeks. Even so, 'the reality was quite different from the expectation'.

Separating them from the house there lay a sunken road and some wire. 'We lay down as close as we could to the wire and then flung two volleys of hand grenades into the balcony of the house.' Then they charged, shouting 'Cabar Feidh!' – Gaelic for a stag's antlers and the call of the Seaforth Highlanders. And, suddenly in execution, the planning felt – as all good planning should – completely over the top. 'If Jerry had had the wit or the guts to put up a good resistance he must have done us great damage ... He did not.' Charteris had arrived just in time, and the escape route down to the water was cleared. There was now just one last element that could go wrong: the extraction from the beach by boat.

To work, the mission required coordinating land, air and sea. It was a collaboration between the RAF, for the drop, the Army, for the assault, and the Navy, for the escape. This meant that it also required conditions to be right for all of them: weather for the planes, moonlight for the troops, and tides for the boats.

In Naval headquarters in Portsmouth, they had watched the weather nervously. A year earlier, Admiralty House had been hit by the Luftwaffe, so Admiral William James, Commander-in-Chief, Portsmouth, had taken what he considered the most practical alternative accommodation befitting a Portsmouth naval officer. He had moved into HMS *Victory*, Nelson's dry-docked flagship. He took up residence in Hardy's cabin, while his wife used Nelson's as a sitting-room.

For days, as the Navy stood ready for the extraction, the weather had been deemed too bad and the mission had been postponed. They had waited so long that the mission window – the period when the tides aligned to make extraction possible – had passed. It had looked like they would have to wait another month. But then, on the morning of 27 February 1942, just as the weather did improve, Frederick Cook, the Australian commander in charge of the Naval extraction, presented a picture postcard to James, and he decided to take a punt.

After the British were pushed out of France, a call had gone up for people to send holiday photographs – to reconstruct in detail the French coast in preparation for the return. There was a particular interest in those that showed people paddling. From the distance out the bathers had gone and the height the water reached it was possible to reconstruct the gradient of the beach. In this case, the bathers in the Bruneval postcard presented to James were enjoying a holiday on an unfortified, idyllic beach. Crucially, they were barely in the water but already up to their waists. It was the sort of steeply shelved beach that could take a landing craft even with an unfavourable tide.

It was enough. Admiral James sent a message: 'Proceed with Operation Biting Tonight, 27th Feb.' This dispatch lacked, perhaps, the rousing nature of the message dictated on another fateful day from the same poop deck: Nelson's 'England expects that every man will do his duty.' Mountbatten, with a better sense of history, had already sent his own encouragement to commanders. 'Best of luck,' he said. 'Bite them hard.'

Out at sea, the news that the operation was on had been greeted with jubilation. Steaming out on a clear blue day, those set to crew the landing craft had corked up their faces in anticipation, then, wrote a Reuters reporter embedded with the Navy, 'paraded the ward room in sheer high spirits, giving imitations of well-known black-faced comedians'. But as the appointed hour for boarding the landing craft approached they became more sombre. 'Just before leaving, echoing through the ships came the stirring melody "Land of my Fathers" sung by the Welshmen who formed a large part of the sol-diers' protection crews.' Silhouetted on the moonlit sea, leaving their mother ship, the reporter said the extraction teams were 'like a team of huskies on the trail'.

On the beach, Frost struggled to suppress an air of desperation. They had been in enemy territory for two hours, and they were behind schedule. It was now past 2 a.m., and the headlights of Ger-man vehicles had been spotted. Their remaining luck was surely now measured in minutes. They had released flares but their radios didn't seem to work, and peering out into the black waters they could see nothing. Cox looked out to England despairingly, wondering 'if we should be killed or taken prisoner'. Frost prepared for a desperate last stand. Then, from the beach, there came a call. 'Sir! The boats are coming in . . . God bless the ruddy Navy!'

'Materializing out of the darkness,' said Lieutenant Young, 'came the ugly shapes of our landing craft. Never have such ungainly ves-sels looked so beautiful.'

Above them, the cliff top at last swarmed with the long-feared German reinforcements – but, just in time, from below the Bren guns of the boats opened up to cover the extraction.

By 03:15, the last of the landing craft would leave. Six of the para-troopers were missing; they were later captured. Two were dead. The odds in this, the most daring British raid of the war so far, compared favourably to those of a nightly bomber raid. Of the 120 who para-chuted into France that night, 112 men and one Würzburg made it on board. Once on board, they were given extra duffel coats, bully beef, biscuits, condensed milk and, of course, rum.

CHAPTER 16

The Kammhuber Line

'It is right that the German people should smell death at close
quarters. Now they are getting the stench of it'

SUNDAY DISPATCH

2 March 1942

JONES REALLY DIDN'T WANT to be in Uxbridge talking about
magical radar. The previous day, the day after the Bruneval raid, he
had received a call alerting him that the captured Würzburg radar
was on its way to the Air Ministry. Here was, in his words, the 'Bruneval Booty'. This was the point of the raid, the purpose of all the
planning and all the risk: a chance to see German radar technology
up close, to at last see what his enemy was capable of. But instead of
being there to see it arrive, he had a prior engagement he couldn't get
out of. So here he was, at the headquarters of the RAF's Number 11
Group, listening to their theories about how the Germans could now
'see' the bombs in the planes.

One of 11 Group's roles involved sending up sorties to try to draw
out the Luftwaffe fighters. The Fighter Command group would go up
en masse, presenting what on radar looked like the soft underbelly of
a bombing mission. When the Germans arrived, though, they would

instead find it was bait, bristling with Browning machine-guns. Sometimes the RAF set their bait by sending up a few bombers with a hefty escort of fighter planes. Then, the Luftwaffe would bite – flying up to meet the British planes before realizing too late they were outnumbered. Sometimes, though, 11 Group would just send up fighter planes, without a bomber in their centre for them to escort at all. Then, mysteriously, the enemy never engaged. Clearly, the RAF pilots surmised, the German radar could distinguish bombers from fighters and knew when it was being fooled.

Jones sat there listening to this tall tale, feeling itchy, wanting to get back to the Würzburg. He told them that as far as he knew no radar had such a capability, and nor could it ever have. So what was going on? Then he had an idea. He asked them a question. When the fighters went up on their own, pretending to be bombers, did they fly at the speed of a bomber, or a fighter? 'There was,' he recounted, 'a stunned silence.' One of those present shouted 'Christ!' The Luftwaffe could indeed tell they weren't bombers – because they were flying 100mph faster than any bomber could go. This minor conundrum solved, Jones hopped into his car and returned to the Air Ministry 'as quickly as my old Wolseley would allow'.

Even a cursory inspection of the Würzburg by Jones led him to conclude it was 'obviously much better engineered' than its British counterparts. He didn't have time to pause for long, though – his road trip across the Home Counties had only begun. Throwing Frank and the fiddly bits of the Würzburg into the back of the Wolseley, he set off north again – this time for Felkin's interrogation camp, where the hapless Heller, fresh from his excitements of dangling off the Bruneval cliff, was waiting for him. There, sitting on the floor of the Cockfosters Cage, the pair watched as Heller put the pieces together.

Heller was good with his hands but was also, his interrogators concluded, 'of limited intelligence'. He couldn't handle Morse code, and it soon became apparent he had not even the most basic understanding of what radar actually did. 'Even after over two months' practical experience of the Würzburg-Gerät,' they wrote, with evident disappointment in their 'prize', 'he still believes that the instrument "sees" the aircraft in some way and is consequently less effective in bad than in good weather.'

Ever since Heller was discovered on the cliff edge he was destined to become a figure of comedy for British intelligence. In a report on his meeting, Jones noted, drily and perhaps a little pointedly, that they probably only had themselves to blame for ending up with the dunce of the regiment. 'Since he expressed the opinion that his colleagues would have come, if only we had not shot them, we must count ourselves unfortunate that we did not have a more competent selection of operating talent,' he said. 'However, the prisoner's readiness to speak, coupled with his inability to make up false explanations, largely compensated.' Had they had a prisoner with a greater ability, Jones reasoned, his knowledge might well have been offset by a 'corresponding ability to deceive'.

As amusing as Jones and others found him, Heller's life had been one of tragedy. In 1939 he was conscripted into the air force. In September, just after Britain declared war, his newborn daughter became seriously ill and he applied for compassionate leave. It was refused, and he went AWOL. He was found, of course, arrested and imprisoned. When he was released, his daughter had died and he went AWOL again, this time going on a two-day bender. He was sentenced to forced labour.

Finally, on release, he was sent to Bruneval. Heller had only been a free man – in a manner of speaking – for two months by the time the paratroopers arrived. And while they and subsequent interrogators might have derided his level of intelligence, in one aspect he had been remarkably astute. In January he noted to his wife that the location of the radar site was so remote he feared they might be snatched by English paratroopers. He was now wondering, Felkin said, whether his wife 'was a fifth columnist'.

Heller's experiences of the war so far probably explained his willingness to talk. Having been imprisoned for twenty months, having been denied the chance to be with his daughter when she died, Heller felt little loyalty to or affection for the Nazi regime.

All afternoon Jones and Frank sat, watching him slot together the pieces with evident competence, hearing him describe life on the

Würzburg station. 'He spoke quite freely,' said Jones, 'save for occasional intervals devoted to violent efforts of memory in order to recall some simple point.' Listening to him, Jones concluded that the equipment is 'probably capable of good technical performance'. But, he added cuttingly, 'in the hands of an operator such as the prisoner, its functions would be seriously degraded.'

The mismatch between the quality of the equipment and the quality of the personnel was not entirely coincidental. After the war, Jones had the opportunity to meet General Wolfgang Martini, head of German Air Signals and Radar. Like other generals responsible for the defence of the Reich, Martini complained that he had to make do with the dregs of the Reich – those not needed on the front line. To compensate, the machinery they operated needed to be, quite literally, fool-proof. Nevertheless, proofing against fools sometimes eluded even German manufacturing protocols – as one anecdote told by Heller attested. 'The prisoner related with joy the fact that the Flak batteries in Le Havre fired blindly at an object reported by him to be high in the sky.' Only after relaying the location did Heller look out of his window and see it was a ship.

This was a fine piece of technology, but clearly not without faults. On the side of it was a tally of kills – of targets in the air and sea destroyed with the assistance of the Bruneval Würzburg. There were fifty-eight of them – painted on with pride by Heller's colleagues.

'One fact,' writes Jones, 'betrays the falsity of these claims. On 5/12/41, three aircraft are inscribed. Now we have excellent reason to remember that day, because it was overcast with low cloud, and there were practically no operations by our aircraft. It was, however, a good day for low obliques and was in fact the day on which Squadron Leader Hill took the low obliques of this very station, which led to its capture. As his was the only aircraft to go anywhere near the station we can only conclude that the Würzburg signpainter-operator did more signpainting than operating on that day.'

With the captured Würzburg and its captured operator, Jones had filled in a key technological gap in his knowledge. He and his colleagues understood at the level of an individual radar station what the Germans were using to track planes. But what about the wider picture? How did these radar stations coordinate? How many were there? Here, unexpectedly, the Bruneval raid had yet more booty to give.

In the aftermath of the raid, the Germans fretted about the tactical weaknesses it exposed.* Or, as Jones put it, 'The Bruneval raid evidently activated the local defence officer.' Perplexingly, the first security response was to demolish the villa. Then all around the Bruneval site was erected 'an entangled pattern of barbed wire exceeding in complexity anything yet seen'. And, crucially, not only there.

Before, when seen from the reconnaissance planes, the Würzburg radar had been mere specks, flecks of dust on the aerial photographs. Now they all became ringed with fortifications. Beneath the newly installed barbed wire the grass grew. Against its sides rubbish caught and collected. They were specks no more; it was as if each had been circled.

---

* Not only the Germans. For some time, Jones and others had been concerned about the Swanage site where TRE were based. It was close enough to France that there was a possibility its experimental radio work could be picked up by electronic eavesdroppers on the other side of the channel.

As the Bruneval raid had shown, there was the even more troubling possibility that it could be raided. Jones wanted it moved. The problem was that its occupants, who had rather a nice life on the South coast, did not want to move.

After the raid, Jones visited. Edgily fingering his revolver, he 'contrived to give the impression of nervousness and an anxiety to return to London as soon as possible'. He made it very obvious that he feared assault by paratroopers at any moment.

Jones was, it transpired, just one source of pressure. Albert Rowe, one of the radar scientists stationed there, complained that all sorts of rumours were swilling around. 'There were, we were told, seventeen train-loads of German parachute troops on the other side of the English Channel.' Churchill, he was told, wanted them out before the full moon. An infantry regiment was coming, he was warned, to put detonation charges on their equipment. 'My own time was spent less in dealing with the work of TRE than in discussions on whether we should die to the last scientist or run and, if the latter, where.'

They decided to pre-empt the discussion, by running before the paratroop hordes arrived. The next Jones heard, the Swanage team was preparing to move *en masse* to Malvern.

Suddenly all along the Atlantic coast what had been invisible became visible – a great net of radar, ready to catch the British bombers. An even greater intelligence coup was on its way, though. Soon, Jones would be given the key to unravelling it all.

The Brussels–Lyon express departed each day, carrying those few Belgians with the papers that proved they had permission to travel. For 500 miles it steamed south through the eastern side of France. It passed checkpoints and borders, garrisons and Gestapo – but kept on going. This journey began in the temperate grey of flat, northern Europe. It ended in the warmth of the South of France – in a city where, on a clear day, you can see the Alps rising majestically on the horizon.

This city, in 1942, was part of Vichy France. And, often, this particular train carried more than just the functionaries of the Third Reich. In its coal carriage were documents, hidden by the Resistance. Secreted here, they could unobtrusively and efficiently cross the national border with Belgium and the more important administrative one – where Nazi power was nominally ceded to that of Marshal Pétain. And if anyone suspected they were there, they could be shovelled into the furnace and vanish. From Lyon, these documents were smuggled over the Pyrenees into Spain and then, finally, to neutral Portugal – from where they could complete their journey in a diplomatic flight.

One day in the spring of 1942, one of these documents found its way to Broadway, and from there to Jones. It was dishevelled, dogeared and unpromising. He had almost missed it – it had been stored amid a pile of other papers on the shelves of a colleague. But as Jones unfolded it, he caught his breath. Here was a map of the entirety of southern Belgium, showing the position of the night defences. It was a stunning piece of intelligence. 'I like to think,' wrote Jones, 'that one of our Belgian friends, daunted by the prospect of cycling round the countryside laboriously plotting the site of individual searchlights, thought it would be simpler to break into the headquarters of the Regimental Commander and remove his map, for this is what it was.'

The organization of the defences was, he noted, 'rigidly Teutonic', built from a repeating pattern across the country of 'boxes' – each containing two Würzburgs and one Freya. Each box formed a single block in what the RAF was calling the 'Kammhuber Line' – much, apparently, to the gratification of Josef Kammhuber, the rigid Teuton in charge, who was delighted to learn he had joined such luminaries as Siegfried and Maginot in getting his own line. As an opposite number, Jones developed grudging respect for Kammhuber. While it was common practice to exaggerate the effectiveness of your own side, he found that Kammhuber typically reported shooting down the same number of planes as the RAF reported losing. A man who will lie about his own side will often miss his own side's mistakes – and in so doing fail to rectify them. 'I disliked fighting Kammhuber, partly because I did not like fighting an honest man,' Jones said after the war.

The RAF's casualties, though, did not need to be exaggerated. The figures were, over time, grim. In 1941, on every sortie one in forty bombers did not return – with the RAF estimating half lost to flak and half to fighters. In 1942, one in twenty-five bombers were lost on each raid, two-thirds being downed by fighters. This was why working out the mechanism of the line had become such a priority. 'New boxes began to spread ivy-fashion over the whole of Western Europe,' wrote Jones. What went on in them? It was increasingly crucial to find out. 'Although the losses in each box would rarely be high, they would rarely be nothing; in aggregate they would mount up to a total never spectacularly high, but never spectacularly low; and over a long period they might well prove crippling.'

While happy to adopt the British name for the line, Kammhuber preferred to refer to the boxes as 'Himmelbetts' – four-poster beds. Each was roughly twenty miles long in the north-south direction and thirteen miles wide, and contained in its centre, usually within a few hundred yards of each other, one Freya and two Würzburgs.

By the time Jones began seriously to tackle the problem of Kammhuber's line, his opponent had lost his searchlights. They had been pulled back to provide a visible defence for Germany's cities. Kammhuber had also found, though, that he didn't need them. Radar meant his planes could see in the dark. Thanks in large part to the agents of the Belgian Resistance, Jones was beginning to work out how.

Each report sent back was testament to men, women, girls and boys prepared to go to their deaths for their country. The utilitarian technical descriptions also spoke of the quiet bravery of a cyclist on a moonlit street, of a patriot creeping through hedgerows to get that bit closer – knowing that capture meant torture, betrayal and execution.

Jones never knew how the map of the Belgian defences was obtained. He never learnt the details of the heroism, trickery or cunning that resulted in it finding its way to a dusty shelf in his dusty building. Neither did the man or woman who found it know how it was used, or even if it was seen at all. He or she could see, though, that the radar sites they so diligently reported stayed unmolested – still spinning, still detecting planes.

With one, later, microfilm description of a radio installation, Jones received a note – from an agent codenamed VNAR 2. 'In view of the apparent current interest in similar installations, it seems odd to us that no attempts have yet been made to destroy them.' The agents had no doubt of their value. 'The Germans' interest in them is clearly shown by the extremely strict way in which they guard the approaches, which has several times resulted in our being fired at by sentries, fortunately with more zeal than accuracy.

'As far as our work is concerned, it would be helpful if we knew to what extent you and the British services are interested. We have been working so long in the dark that any reaction from London about our work would be welcome to such obscure workers as ourselves. We hope this will not be resented since, whatever may happen you can rely on our entire devotion and on the sacrifice of our lives.'

There was no way to let the agent know that the work was not in vain. Doing so would have constituted unnecessary communication, and unnecessary communication was unnecessarily dangerous. The RAF was, in fact, far more interested in understanding the radar defences and being able to defeat them *en masse*, than attempting to take them out piecemeal. The policy at the time was to leave counter-measures until as late as possible – rather than reveal their hand.

True to VNAR 2's word, though, the intelligence continued to flow. Agents continued to send the position of radar stations. One, a Brussels jeweller, bribed an SS guard to get inside a control station, where he spent several nights sketching the equipment. It was, Jones said, the most detailed piece of intelligence they had on nightfighter control until after D-Day.

So it was that between the sketches, the agent reports, the Enigma intercepts and the photographic intelligence, towards the end of 1942 Jones declared that they had fleshed out the entire defences, and knew how they worked.

Imagine a raid. Not a big one, nor a small one, just one of the many routine bombing runs of 1942; with their routine triumphs and tragedy. It is a late spring day, and dusk is settling over the English Channel. From airfields across the south of England, the bombers assemble, and prepare to breach the invisible Kammhuber Line. Now imagine a box in that line. There is nothing remarkable about this box either, it is just a brick in the wall defending the Reich, identical to those either side. Except, tonight, this unremarkable box is where this unremarkable bombing run is set to cross.

The first warning comes from the Freya. Peering 100 kilometres and more towards the south coast of England, it picks up the first reflection from the leading bomber.

Circling around a radio beacon in the dark is a waiting nightfighter. This is the nightfighter assigned to intercept anything passing through this box, and it is told to be ready. In the plotting station, from where the battle will be directed, a dozen staff assemble. Some arrange themselves in a gallery around a large, round, glass table. Others are at floor level. The table, suspended above the floor, is etched with a map of the coastline and, centred on the middle, a red circle. On the scale of the map, the circle has a radius of 36 kilometres – the range of a Würzburg. When the bombers breach that circle, the circle that means their height can be determined, battle commences.

In the plotting room, three staff are in telephone contact with the Würzburg – one for each dimension. One passes on the altitude, one

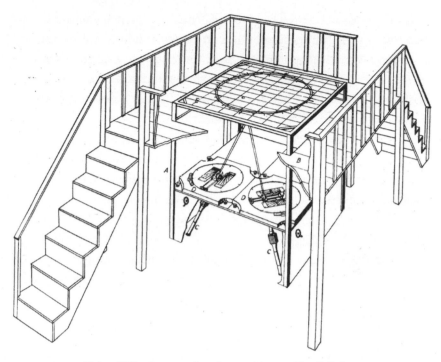

*Using Würzburg radar the positions of both fighter
and bomber could be plotted on a single map*

the bearing and one the range. Together, they control a single red light projected on to the table, giving the position of the bomber. This is joined by a second light, blue this time – its position provided by the second Würzburg. It marks the nightfighter. The job of the men and women in the plotting room is to bring those two lights together, by speaking directly to the nightfighter pilot.

Jones was somewhat scathing about this process, regarding its planners as typically technocratic Teutons. 'It appears to be a sub-conscious point of German policy never if possible to leave the decision to the man,' he said. Instead, 'it was a natural concept for them to regard the nightfighter as a ground-directed projectile.'

But it works. On this particular night, the lead bomber is picked up in the glare of the red Würzburg. Although it has no way at this stage of knowing this, its own particular odds are the spin of a Russian roulette. Once locked on, one time in five, the red dot will merge with the blue dot: the fighter will have found its target. Generally, this happens within ten minutes. Often, it will only be the bullets ripping through the plane that alerts the bomber to the danger. The shrewder nightfighter pilots aimed to attack from behind and below, getting to within 20 to 50 metres before opening fire. If they made contact then, 60 per cent of the time they were successful.

On the plotting table, a draughtsman with a chinagraph pencil traces its route, and marks a kill. The Würzburg continues to follow the bomber as it drops in flames – when it disappears, its last known location is called in so that a staff car can be dispatched to the wreckage. Then the fighter returns to the beacon and the process begins again.

There was a brutal, mechanical efficiency to the system. And yet, there was also a very obvious flaw. As Luftwaffe general Josef Schmid put it, the entire box was 'an excrescence of the control of one fighter'. All this effort, all these people, were there to manage a single man on a single target. Like the crocodiles of the Mara River, awaiting the annual wildebeest migration, there was only so much prey that could be taken down by this method. As with the wildebeest, some of the cannier Allied bomber pilots had already worked out the implications of this – wait until the crocodile is engaged, and then rush through. 'Some pilots would take advantage of a time when they could see some unfortunate colleague receiving the individual attention of the ground defences to slip into a target area,' noted Jones.

This idea would be formalized in a simple countermeasure. Spread out the bomber force, and the nightfighters of each box would have plenty of time to pick off targets. Concentrate it, though, so that it passed through a narrow region rapidly, and they were far more limited. The concept of the 'bomber stream' was born. But simply overwhelming the defences still meant accepting that, each night, like the hungry crocodiles, the nightfighters would take their lunch. It was also becoming clear that the German planes were getting better at doing so, that some no longer needed to rely only on the radar from the ground. The Luftwaffe were not out of tricks yet.

*

Forty years later, when both of them were old men, Professor Reginald Jones CBE would send Sir Charles Frank OBE a letter. 'Dear Charles,' it began. 'Do you recall the message from Belgium . . . ?'

Jones had, he said, at last discovered the identity of agent VNAR 2, the source who sent them such a plaintive note. He was called André Mathy, and he was a doctor who, it turned out, was almost exactly the same age as them.

In the four decades since, Frank and Jones had had full lives. They had gone on to have distinguished academic careers in their own right at opposite ends of the country – Jones in Aberdeen, Frank in Bristol. Both had become fellows of the Royal Society, Britain's most venerable scientific institution.

Mathy was captured on 13 March 1943. On 21 June 1944, two weeks after the Allies landed on French soil and the liberation of Europe began, he was executed in the Belgian city of Halle.

CHAPTER 17

Radar Goes Mobile

'The pursuit of the good and evil are now linked'
SIR BERNARD LOVELL, FOURTH REITH LECTURE, 1959

10 July 1942

AT ABOUT THE TIME that Jones was perfecting his understanding of the German radar, another young radio scientist, Bernard Lovell, was creating a new kind of radar entirely. Or, at least, he was trying to. But it was not, in his view, close to being ready. Which was somewhat awkward, because Winston Churchill wanted an update.

Lovell had been invited – ordered might be a more accurate description – to Downing Street to report on this new kind of radar: one that provided an image of the ground below. This radar held out the promise that, at last, British bombers could find and hit their targets, whatever the weather. So impressive were early results that Albert Rowe, a senior government radar scientist, had declared, 'This is the turning point of the war!'

Even so, to Lovell, right at that particular moment his research seemed an odd priority for the Prime Minister. On the one hand, Egypt looked set to fall. On the other hand, he and his colleagues' work on making the radar a practical reality had until a few weeks

ago been going pretty badly – at which point it started going cata-
strophically: the only prototype got destroyed in a plane crash, along
with eleven of Lovell's colleagues, including Alan Blumlein, one of
the greatest radio scientists of the age. Rowe's turning point was
proving further and further off. Yet, despite all this, here they were.

Churchill arrived, late, wearing a blue boiler suit zipped at the
front. He pointed to everyone in turn, demanding their names. He
seemed distracted. Crossly, four times the Prime Minister inter-
rupted the meeting to buzz through to his secretary for his briefing
notes. Four times he received the reply that he already had them.
Each time, he also demanded that someone was sent to take minutes.
Eventually, a cowed John Llewellin, Minister of Aircraft Production,
took on the task – later to be relieved by a major-general.

Luckily for the major-general's shorthand, there weren't that many
minutes to take – Churchill was already late for a cabinet meeting
and wanted to rush through the proceedings. He listened to an
update on the research, chewed three cigars, threw them over his
shoulder, and then stated that by 15 October there were to be two
hundred working sets of the radar, of which currently there were
zero. Despite the scientists' objections, 'no one could shift him', wrote
Lovell. So, dejectedly, they 'traipsed out', having 'a close-up view of
the moth-eaten cabinet' who were, given Churchill's mood, presumably also about to get it in the neck, and then went to a room next
door to decide among themselves 'how to do the impossible'. When
he got home that night he recalled in his diary that the entire day had
been 'an unbelievable pantomime'.

If Lovell and his fellow scientists were, indeed, to do the impossible then their task would begin with the Allies' newest invention,
perhaps the most important invention in radar so far. It was an
invention that at last promised that radar could be fitted on planes
themselves. The problem with radar is that it is big. It is big because
of all the equipment it needs, but it is also big because of a fundamental limitation. The aerial needs to be in proportion to the
wavelength. If you want a miniature radar, it will project extremely
short waves, and you will have to find a way to pack enough energy
into that miniature-sized apparatus to still project them. Miniature

radar tended to have miniature range. But the British had an advantage: the cavity magnetron.

Looked at side-on, the cavity magnetron isn't much. It is a tube of metal hollowed out with chambers, a little like a revolver, but with a solid rod where the gun barrel would be. Its effect, though, was to revolutionize electronics. It was so significant that a top-secret British mission was sent across the Atlantic to present one to the Americans, like a medieval vassal buying the protection of a more powerful ally. In 1946, a US historian described this delivery as 'the most valuable cargo ever brought to our shores'. Even today you still have something like it at home, in your microwave oven.

Like conventional radar devices, the cavity magnetron produces electromagnetic waves by moving electrons. Its trick, though, is to amplify them. Think of it as a little like a wind instrument. If you blow into the air, you make a faint sound. If you blow through a flute, causing your breath to resonate inside the cavity – you make a loud sound. A cavity magnetron is a flute for electrons. As they whizz past the cavities, they resonate. Then from out of these cavities blasts a high-intensity beam of electromagnetic waves. These waves were not the same as those Britain had used before. They were packed closer, their shorter 'centimetric' wavelengths allowing a tighter beam.

The point about the cavity magnetron was that it was very compact and very powerful. It was a way of miniaturizing radar, so that you could, for instance, take it on a plane. The first, most obvious application of the cavity magnetron would have been to put it in a plane – to give that plane its own miniature version of the land-based radar – and this had indeed been Lovell's initial assignment. Radar in Britain had begun as vast edifices, hundreds of feet high, to direct planes to their target. What, though, if the planes could direct themselves? Could he make a device that allowed nightfighters to home in on the enemy planes from the air?

For Lovell, the shift from lab work to practical military research had been a learning experience. 'Hitherto in my few years of university research I had donned a lab coat and dealt with an apparatus in the warmth and quiet of a laboratory,' he wrote afterwards. 'Now the lab coat was replaced by a bulky flying suit, a parachute harness and a parachute which also served as a seat. The spacious laboratory was replaced by the cramped and cold interior of a Blenheim night-fighter. The noise made normal conversation impossible and the vibration was so great that I could not imagine how any electronic equipment could survive.'

The work was initially conducted on the south coast – until the success of the Bruneval raid led to sufficient fears about their own vulnerability for the laboratory to be moved inland. It would prove to be the basis for the first British airborne interception radars, which would at last tip the balance in the skies above Britain. And yet, along the way, there were a few hints that it could do so much more. One hint had come in the summer before, when one afternoon, essentially on a whim, Lovell had asked an assistant to cycle along the edge of the sea cliff near their Dorset base, holding a large tin sheet, the flat side pointing towards Lovell. Using their prototype airborne radar, albeit a grounded version, they could see a clear reflection bouncing off it. As Reg Batt, the assistant in question, continued his precarious bicycle ride, presumably hoping the wind didn't pick up and turn the sheet into a sail that would topple him, the path took him along a route where the land between him and the cliff rose up, so that between Batt and the sea was a solid bank of earth. At this point, he should have disappeared – the signal obscured by the earth.

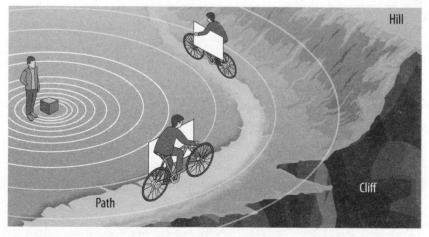

The cyclist showed up clearly against the hill – showing that radar could differentiate between surfaces

And yet, Lovell noticed, the strength of the echo was 'amazing'. The land behind was still sending off a weak signal, but it was so much stronger from the tin sheet. Clearly, radar differentiated between different surfaces and different materials.

Another hint of what this radar could achieve began as an annoyance. One of the problems with airborne radar is that the ground interferes with it. As a radar beam spreads, it begins to bounce off the land below, these echoes mixing with the signal. On the screen, all you want to see is the clear blip of a plane ahead. Instead, radar waves return from below, making the clear blip fuzzy. This is another reason why the tight beam of a cavity magnetron – which could travel further before it hit the ground – was preferable. Even so, in early tests, crew said they could see the coastlines appear on their radar. From Leeson House, where the radar research was carried out, the town of Swanage below was clearly visible on the radar.

None of this should have been surprising. Radio waves are just light outside of the visible spectrum. When visible light shines down from the Sun on the Earth, it is reflected and refracted, absorbed and scattered. The result of all that light bouncing off the ground in all those different ways is the world we see around us – colours, textures and shapes. Why shouldn't radar do the same? Why shouldn't it show textures? Why shouldn't it differentiate between the reflective surface of a factory and the duller, more muffled return from the field either side? What if, far from getting in the way, these reflections are useful information – information that shows you the shape of Berlin on a cloudy night, that reveals Kiel harbour in a blackout? That shows you the Rhine even when the moon has set? Then, you have something of real value – something that might just, in the words of Rowe, be a 'turning point'. You have, on your radar screen, a map of the ground below.

Churchill did eventually get his two hundred sets of what became known as H2S,* but not by October. By the end of the year, though,

---

* Quite why it was given that name is a matter of some historical controversy. One idea is that it was shorthand for 'home sweet home', because it homed in on homes. Jones, who would retain scepticism about the technology, claimed

there were twenty-four planes equipped with it, waiting for the order to go.

Still, Lovell waited nervously. There was an odd stipulation about its use. As development had continued, in the Air Ministry there had been a debate about whether its use on bombing missions was sensible. If a plane crashed and was captured, the Germans might discover the secret of the cavity magnetron. There had been attempts to add an explosive charge, to sabotage it if necessary, but at the core of the cavity magnetron was a large moulded copper block that was almost indestructible.

Eventually, a compromise was reached. If the Russians held the Volga, and the war appeared to be turning, they would risk H2S missions over enemy territory. For weeks, Lovell watched the eastern campaign anxiously. In a series of desperate battles, in a terrible winter, at great cost the Red Army threw the Germans back from the outskirts of Stalingrad. On 30 January, in Berlin, Goebbels read out a speech praising the 'heroic struggle of our soldiers on the Volga'. The heroic soldiers, surrounded, frozen and starving, were on the point of collapse. In response to warnings from Friedrich Paulus, the German commanding officer, that all was lost, Hitler promoted him to Generalfeldmarschall – a move that would seem quixotic but for the fact that, in doing so, he pointedly noted that no German field marshal had ever surrendered. It was a signal to fight to the end.

The battle was, indeed, lost. That same night, just before Paulus – in defiance of German military tradition – surrendered, H2S flew.

Here, on the Western Front, on the other side of the German empire, it was a bad night for bombing. The pilots noted that there was no moon. Over the coast, there was cloud to 20,000 feet. Over the target there was a lot of haze, and essentially zero visibility. But, at last, it didn't matter. Watching the gentle glow of their monitor, the crew had seen through to the ground in dazzling clarity. They had

it was because their delays in getting it going had led Lindemann to declare 'It stinks! It stinks!' and H2S is the chemical formula for hydrogen sulphide – which smells of rotten eggs.

identified all the landmarks that would have helped them had it been a daylight mission in clear weather – Heligoland, Zwolle, Bremen, Cuxhaven and then Hamburg itself. Towns appeared when they were still twenty to thirty miles away.

Group Captain Dudley Saward, chief radar officer in Bomber Command, debriefed the navigators. 'Coastlines, estuaries and rivers were described as appearing on the cathode ray screen "like a well-defined picture of a map",' he wrote. The navigators who reached the target 'claimed positive identification of the docks, stating that the coastline appeared as "fingers of bright light sticking out into the darkness of the Elbe".'

As Britain at last went on the radio offensive, the system was just one of the navigational innovations to find its way, at last, into the RAF. It would still be in use operationally during the Falklands War. The other innovations, developed in parallel, were a lot closer to those already used in the German navigational beams, and, also, thanks to an intelligence screw-up with the potential to leak the details to the enemy, very nearly suffered the same fate.

H2S was just one part of a rapidly evolving military doctrine in Britain. To have land supremacy required air supremacy. To have air supremacy required, it was now clear, airwave supremacy. The days of 1939, when bombers were dispatched like ships of the night, navigating alone by the stars, seemed like a barely imaginable world – like looking back on the cavalry charges of 1914 from the trenches of 1917. Just as in the trenches, they needed to push forward, gaining territory in the sky.

First, Jones had denied the Luftwaffe the use of the electromagnetic spectrum over Britain – blocking and distorting the beams. Now, for the war ahead, the Allies needed to gain control – or at the very least contest – the electromagnetic spectrum over occupied Europe.

If H2S gave bomber crews a map, a system codenamed Gee put gridlines on it. Imagine you have two radio towers, evenly spaced, called A and B. These towers, in this example, are 300km apart, sending out the same radio signal.* Even though the signals are sent at the same time, as far as a plane is concerned they will reach it at different times. The one from the closer tower will arrive sooner. The difference between them tells you how much longer the radio wave is taking to reach you from one station compared to the other. If, for instance, the signal has taken half a millisecond longer to travel from B than A, then that tells you that you are 150km closer to A than B, because light travels 150km in half a millisecond.

This on its own is only of marginal use. There are an infinite number of places where this is true. You could be 500km from B and

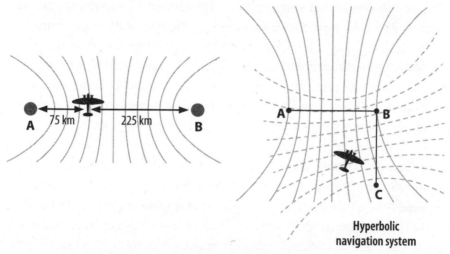

Gee created grid squares in the sky

350km from A. You could be 225km from B and 75km from A. You couldn't, though, be right on top of A and 150km from B – for the simple reason that the sites are 300km apart. So knowing that the

* This sounds like a difficult problem to synchronize. How do you ensure they transmit at exactly the same time? Actually, it's not so hard if one of the stations acts as a slave to the other. Station A sends the signal, station B receives it. Because it is, in this case, exactly 300km away and we know light travels 300km in 1 millisecond, it then re-radiates the signal, but just corrected by the known time difference.

signal is offset by half a millisecond tells you that you are somewhere on a curved line that stretches towards the horizon in both directions.

That is one set of grid lines. Make another, running perpendicular, and suddenly – just as on an Ordnance Survey map – you have a reference system. You have a grid.

That was the idea. There was, though, a problem. The RAF, an exasperated Jones complained, had made exactly the same mistake the Germans had made two years earlier. Before introducing the system, it had tested it out with a bunch of airmen, some of whom had then ended up shot down – one crew on a plane containing the apparatus itself. The RAF now had to assume they had been interrogated in the German version of the Cockfosters Cage, and some had given away the secret – either directly or through subsequent eavesdropping.

'It therefore appeared probable,' wrote Jones, 'that if the German Intelligence were alert, they had in the receiver and prisoners enough data to unravel the Gee system.' Was the game already up?

There was nothing they could do about the lost plane and the crews. All the Nazis had to do was join the dots. But the thing about joining dots, as anyone knows who has pondered how our ancestors could look at the stars and see a bear or a great dipper, is that there are lots of ways to do it. What if, Jones reasoned, they accepted that the Germans already had the key clues – but then provided them with other dots to which they could join them, making an entirely false picture?

He recalled their own detective work in unravelling the mystery of the German beams. 'Knowing the types of clues – often slender – on which we had to depend, it was fascinating to see whether we could present the Germans with similar, but false clues. In order to ensure this it was necessary to decide upon a completely false but plausible picture of our own activity, to which they would be allowed to gain clues.' This false picture they would call 'Jay'. It would be easy enough for his opposite numbers to convince themselves that either they, or the prisoners, had misheard 'Gee', and if the Germans did hear more about the true scheme, the phonetic similarity would merely heighten the confusion.

So it was that, early in 1942, German intelligence received a report. Their agent – an agent actually controlled by the British – told them he had overheard a conversation between two RAF officers in the Savoy. One of the officers was cross. A colleague had received a knighthood and, frankly, he didn't think he deserved it. 'All he has done is to copy the German beams, and a year late at that! In any case, it wasn't him alone, but the chaps under him.'

His companion was emollient. 'You must admit that at any rate we now have the "J" beams to get us to our targets; they worked OK on Brest, and we shall soon have them over Germany.'

'An argument followed,' Jones explained, 'in which Brest and the relevant "J" system were marked with a fork on the tablecloth.' Jones confessed to being especially proud of the cutlery. That had added a real touch of verisimilitude.

This conversation, of course, had never happened. Most of the German agents in the UK had been captured, turned, and used to send back misinformation. This particular misinformation scheme was just one part of what Jones called 'a marvellous opportunity' to apply 'as much of the national resources at my disposal as I wished' to 'practical joking'.

Jay was specifically designed to flatter the Germans. The Jay beams were to all intents and purposes a replica of the Luftwaffe beams. 'To lend verisimilitude to the beam story,' Jones explained, a real beam station was used – pumping out real beams to be heard by the German listening service. 'Almost every step taken to suppress Gee was followed by a similar step to conjure up Jay,' explained Jones.

The plan was not to prevent the Germans finding out about the system forever. 'It was too much to hope that we should continue to mislead them once serious operations had begun.' If, though, the RAF could at least begin those operations without giving the Germans a head start, that was enough. The hope was there would be perhaps three months of unimpeded use of Gee. In the event, after it finally came into full service later in 1942, there were five months.

In his post-war debrief General Josef Schmid still referred –
apparently wholly unaware of the dupe – to 'Jay beams'. Although,
when he did, oddly he seemed to be referring to yet another system
entirely – a system that was in fact a little closer to the beams first
used by the Luftwaffe. Because while work was continuing on developing Gee, in parallel there was a second navigation device called
Oboe. Whereas Gee could give every bomber over Europe a reasonable sense of its position, Oboe could direct specific bombers on to
targets with pinpoint accuracy. This device was going to allow Jones
to, at last, repay some old debts.

To fix a single position on a map does not require laboriously designing a pair of pulsing beams that cross. In fact, it just requires two
numbers. Let us say that you choose two places – Trimingham in
Norfolk and Walmer in Kent, perhaps. Let us also say that you assign
a distance to each: 220 miles for Trimingham, 155 miles for Walmer.
Now it is a simple fact of geometry that there are two locations in the
world, and only two, that are exactly 220 miles from Trimingham
and 155 miles from Walmer. Since those towns align roughly north–
south, one of those places is east of them, the other is west. This may
be a mathematical truism, but its consequences are profound. If a
bomber can know its distance from those two fixed points, and be
competent enough when taking off to head east for Europe rather
than west for Swindon, then it can always know where it is.

This is what Oboe did. It was a little like a cross between the
Y-Gerät and Knickebein. Like the Y-Gerät, it used a radio signal to
set a distance. From one transmitter in Walmer, codenamed 'Cat', a
signal was sent out. This would then be pinged back by the plane.
From the time taken for the signal to return, the radio operator could
tell the plane its distance. Unlike the Y-Gerät, the plane's job was to
reach a set distance from the cat transmitter and then turn so that it
stayed at that distance – flying in a shallow arc with that exact radius.

This was where a second transmitter, codenamed 'Mouse', came
in. It was positioned elsewhere in Britain, in Trimingham – in this
case 220 miles from the target. It worked in exactly the same way. As
the plane flew along the 155-mile radius circle centred on Walmer,

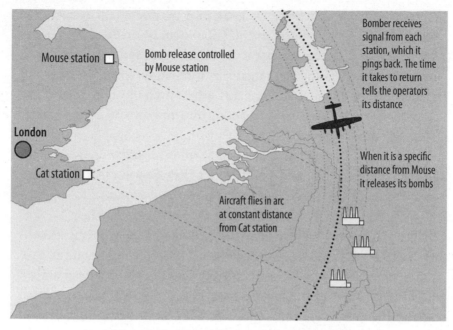

Oboe allowed precise control of a single plane

the operator at the second transmitter would wait until the point when the plane was exactly 220 miles away from Trimingham. This was the intersection point that precisely determined the target. When the plane crossed it, the bombs dropped.

TRE made up a song to describe the process.

> Flickery-trickery chase,
> The Mouse ran down the trace,
> The Cat's away,
> So Mouse can't say,
> Whether pilot or puss sets the pace.

In theory, the system was highly accurate. But they needed to test it out in practice. In late 1942 a series of raids on different targets were planned. Francis Jones, a radar scientist at the Telecommunications Research Establishment, approached his namesake Reginald, and asked for help. Could he come up not just with a German target, but also a team on the ground who could assess the aftermath?

For Reginald, there was one resistance network he trusted above all others.* He told Francis that if TRE would be satisfied with a target in Belgium, he could guarantee the information they needed. A message was sent, and the target was chosen: a German air control centre in the grounds of Florennes Castle, south of Brussels near the border with France. It was a position 220 miles from Trimingham, 155 miles from Walmer.

In late December a small force of Oboe-equipped Mosquitoes headed out from England to hit it. As they flew into Belgium, the home transmitter listened for their pings – and signalled to let them know if they were too far, too near, or just right. Then, in a manoeuvre that would come to be repeated many times by the new breed of 'Pathfinder', the Mosquitoes followed a long shallow arc of 155-mile radius that, they knew, would take them over the target. They kept flying until they received the second set of pings that would tell them it was below. Then, without needing to recognize the ground below or the stars above, they released.

Jones's faith in the Belgians was more than justified. The craters had barely cooled by the time, next morning, a group of resistance fighters arrived. Stepping carefully amid the bomb damage from the night before, they methodically paced out the distance of each hole

---

* In one way or another, over the course of the war he would end up having contact with most – either through the radar war or his other duties. At around this time, he was also trying to understand the intricacies of the German nuclear weapons programme, which centred on a factory in Norway that was understood to be making heavy water, a key step in the process. Working through the Norwegian networks he asked for more information, and was eventually directed to a local expert who had been manager of a paint company. This expert sent the response, which amused Jones greatly, that they would help so long as no commercially sensitive information was passed to ICI, 'Because blood is thicker even than heavy water'.

It was a rare moment of levity. The first raid on the plant involved the loss of all involved. But it was still essential it was destroyed. 'I remember . . . being asked to take the unpleasant decision of insisting that a second raid be mounted,' Jones recalled. It was hugely successful and became the most famous action of Norwegian resistance in the war: a daring operation in the bitter Norwegian winter that was immortalized in the film *Heroes of Telemark*.

from the German air command post. Then, recording their findings in yards – using Imperial in deference to their handlers – they sent the information to Britain.

No one can know the emotions of these unknown resistance fighters as, at considerable risk, they strode among the debris of the attack. Theirs was a shadowy war of terrible stakes. The death rate among resistance fighters was as high as those seen in any unit anywhere in the world's armies, in fact higher, but without the medals and regimental records. Necessarily. It was a secret war, with secret sorrows, in which information normally passed just one way – from the Resistance in Belgium to the intelligence officers in London.

But this time it was different. The bombs falling from above were a reply, a confirmation at last that their intelligence was being used.

Back in London, Reginald Jones fondly imagined that in this group of Belgians that morning there was at last a sense of vindication. Because in ordering the raid and telling them in advance, the information had for once gone the other way. The message he sent from London was orders for a mission, but it was something else: it was a thank you, a way of saying that they were appreciated, recognized and not alone.

Jones imagined too that there was not a little joy when the distance for one of the bomb craters required no pacing at all. It was a direct hit on the Nazi target.

CHAPTER 18

'Emil-Emil'

'The crew fly on with no thought that they are in motion. Like
night over the sea, they are very far from the earth, from
towns, from trees. The motors fill the lighted chamber with a
quiver that changes its substance. The clock ticks on. The
dials, the radio lamps, the various hands and needles go
through their invisible alchemy'

ANTOINE DE SAINT-EXUPÉRY, *WIND, SAND AND STARS*

3 December 1942

THE FIRST SIGNAL APPEARED west of Mainz. It was just the faintest, ominous ping of a radio wave in the frosty December night.
Inside the Wellington bomber, Pilot Officer Harold Jordan, a school-
teacher in peacetime, noted the frequency and warned the crew. But
that was all he did; his bomber kept on flying straight. Over the next
ten minutes the signal became stronger, more insistent, more menacing, more certain. Amid the dark of central Germany's hostile skies
it was like the clicks of a bat homing in on a moth. But still the
bomber flew on, prey awaiting a predator.

Soon the signal inside the cabin was so loud that it overwhelmed
the receiver apparatus. There was no doubt that it was a nightfighter,

nor that it was using radar. Attack was seconds away. Jordan rushed the findings to Flight Sergeant Bigoray, the wireless operator, who scrambled to transmit them in time. That was when the first shell smashed into the side of the plane.

The crew of that Wellington that night were bait. When they had taken off they knew that there were only two plausible outcomes of their flight – a flight requested by Jones. Either they would be unsuccessful, and the Germans would not rise to the bait. Or, they would be successful, and it was, surely, a suicide mission. It appeared they had been successful.

A few months after the Bruneval raid, chatter on the German radio waves had begun to come in of a device pilots called 'Emil-Emil'. 'I have got the enemy on Emil-Emil,' reported one pilot mid-fight. Later, a ground controller asked a pilot to confirm he had seen the bombers on Emil-Emil. By October, Jones was receiving so many reports that he could not keep up. Some also referred to a different word, apparently for the same device: Lichtenstein. He had no doubt that they were, at last, dealing with airborne radar.

It had always been as clear to the Germans as the English that radar needed to go airborne. Directing a plane to its target by speech using ground radar was still fighting blind – just with someone describing over the radio what the blind person can't see. Frequently the German radar controllers would bring the two dots together, only to have the pilot claim to see nothing. Either the altitude was wrong, or it was too dark. Sometimes, the pilot would get in position just as his prey left Würzburg range – and he had to give up.

In 1940, the German radar pioneer Wilhelm Runge had collared Wilhelm Messerschmidt, designer of the eponymous fighters, to state the obvious. 'I ask you,' he said, 'surely the essentials of a night-fighter consist of an eye and a gun? If it can't see, it might just as well stay on the ground. Ergo, you must find a place for the eye!'

The Luftwaffe kept trying, but could find no way to fit their radar inside the fuselage. They had no equivalent of the cavity magnetron, and the idea of putting it outside, of adding an ungainly aerial to their sleek fighters, was not merely unpalatable from an aerodynamic point of view – put simply, it offended their aesthetic sense. Eventually, in the battle between aesthetics and exigency, the latter won out. The Luftwaffe might not have had the cavity magnetron, but they still had options. After a year of dithering, a year of rejecting the idea of sullying their planes' noses with what looked like an array of wire coat-hangers, Kammhuber put his foot down and insisted, and the Luftwaffe began designing experimental airborne radar.

By the summer of 1941, it was ready. Oberleutnant Ludwig Becker, the man who a year earlier had been the first to be guided to his target by ground radar, was once again the test pilot. On the night of 8 August, he took off in an Me 110 from Leeuwarden in Holland. Behind him in the cockpit was Sergeant Josef Staub, his face illuminated by the faint glow of a cathode-ray tube – his interface with the Luftwaffe's newest toy. At first, the process was unchanged. From the ground, the radar operators directed the fighter – bringing the red and blue dots together. But when the range closed to under two kilometres, on the very edge of Staub's display there appeared a fuzzy dot. The Wellington bomber was in range of the Lichtenstein, and there in the cockpit they could see it with their new eye.

For Cajus Bekker, the German military historian, there was a tragedy to this moment. 'It meant that the romantic age of flying was past: the crew had become a mutually dependent team of instrument watchers.' For the crew of the Wellington that Staub had in his sights, the tragedy was rather less philosophical. They were shot down, crashing just over the German border.

Through the autumn, more radar devices were installed, and more Wellingtons were to suffer the same fate. For a while, being a nightfighter in the Luftwaffe was the surest method of becoming an 'ace'. Such was the exquisite control the radar gave them that Becker developed a tactic of approaching from below and behind – attacking in a climb, sheltered from return fire, with the whole belly of the bomber in his sights. He would get forty-four kills before being brought down, in a rare daylight sortie.

As, gradually, radar found its way into the nightfighter force, so too, at first gradually and then insistently, it found its way into the consciousness of the RAF. By the autumn of 1942, they could not ignore the references to 'Emil-Emil' – apparently the nickname for the Lichtenstein radar system – or the bombers that all too frequently plummeted in flames afterwards.

Jones realized they needed to establish the existence of German airborne radar definitively, and in doing so pin down its characteristics. Earlier in the war, it had been possible to do both by scouring the wreckage of downed German planes that crashed in England. Now that the Blitz was over, the only German nightfighters that crashed did so on Nazi-occupied territory. The British could not investigate them. So they needed, instead, to hear it in action.

The general policy of the RAF had been to avoid sending up highly trained radio operators to listen in to enemy signals over Germany. There was a pragmatic, if depressing, reason for this. 'Their average life,' one document explained, 'was not long enough for their efficiency to reach a high level.' Instead, the Telecommunications Research Establishment, where Cockburn's boffins made the new radio equipment, and the Y-service, responsible for listening in to enemy signals, had worked on developing automated receivers that could listen out and record signals, and could be installed in planes to do the job of a more expensive, and less expendable, human.

This new Lichtenstein radar was different, though. Here, the German signals the British were seeking were not those of ground radar passively scanning the skies. The British were instead, as Runge put it, looking for the eyes of a flying gun. In order to listen in to the signal, they had to be close to its source, a nightfighter – so close that they were immediately at risk. Jones knew, and the bomber crew on that December night knew, that for the British to do their job successfully they had to put themselves in harm's way.

As the shells ripped open the sides of the British plane, the captain, Pilot Officer Paulton, threw the Wellington into a dive and the rear gunner returned fire. Pilot Officer Jordan – hit in the arm – braced himself in position and wrote a second message to transmit on the radio. The nightfighter returned to the Wellington in attack after attack. Jordan was hit in the head, badly. Throughout, Flight Sergeant Bigoray transmitted the vital Morse message to England, repeating and repeating the details without knowing if they had been received. He continued even after he was hit in the leg, even after the port engine was destroyed, even after both the Wellington's guns were knocked out, even as – miraculously – the attacker gave up and the French coast appeared.

The Wellington limped on towards Britain and Bigoray, now only able to crawl, was pushed out of the ailing plane over Canterbury wearing a parachute and holding a hard copy of the vital message. Then, duty fulfilled, Paulton turned back to sea. The plane was too badly damaged to land; his only option was to ditch it.

Dawn was at last breaking. There, in the freezing December waters, the remaining crew sat on their sinking bomber waiting for rescue – their useless dinghy holed by a Junkers cannon.

Jones was already on his way to work. By the time he arrived at his desk, the calm and precise technical details of the Lichtenstein were safely in British hands. The bland and dispassionate list of numbers and frequencies could not hide the heroism that had brought it there. All, miraculously, survived.

Over the preceding two years, Jones and his team had worked out that the Germans had radar. He had worked out the kinds – chiefly, the Freya and the Würzburg. They had worked out how they operated, coordinating with nightfighters in blocks that each covered part of the line. They had shown how those blocks came together. And now they had unravelled the technical details of the radar carried on the nightfighters themselves.

'The last major gap in our knowledge of German night defences,' wrote Jones, 'was thus closed.' The question now was, what were they going to do with it?

The answer to that question begins ten months earlier, a few weeks before the Bruneval raid, with one of the greatest naval humiliations in British history.

When he left dock in Brest, Vice Admiral Otto Ciliax knew that what he was about to attempt had not been achieved in the last three centuries of naval warfare. He knew that British naval doctrine was predicated on the assumption that it was impossible. He knew that success would, for the same reason, make him a German national hero; but that, equally, the attempt could be his undoing.

On the night of 11 February 1942, his battleships the *Scharnhorst* and the *Gneisenau* left Brest harbour alongside the cruiser *Prinz Eugen*. For months, they had been the target of bombers from southern England; now they were fleeing.

The Royal Navy had been expecting this. What the British hadn't expected was that the Germans would go east – that they would pass through the English Channel rather than go north and around the British Isles. Not since the Anglo-Dutch wars of the 1600s had a hostile ship made it past Dover intact. Here surely – in the home waters of what had been the greatest maritime power the world had seen – Britannia still ruled the waves.

The submarine HMS *Sealion* had been posted unseen outside the port with instructions to radio back if the ships left. But as luck – or bad luck – would have it, she left at 8.35 p.m. to go and recharge her batteries. So it was that all night, unnoticed, the trio of German ships continued east.

The German Navy's operation had been a long time in the planning. As dawn broke, an 'umbrella' of Luftwaffe fighters circled above the flotilla, to defend them from attack. No attack came. The Luftwaffe pilot, Gerhard Krumbholz, remembered being sent out that morning in a Junkers 88. When, two hours into his flight, he spotted three smoke trails heading purposefully east, he could barely believe his colleagues' ships remained unmolested by the feared Royal Navy. 'Majestically and calmly they move through the turbulent sea, surrounded by numerous security vessels, destroyers and minesweepers. Motor torpedo boats hunt protectively on all sides. Often the boats nearly disappear in the heavy breakers. German fighter planes circle over the naval unit heading for the narrow Channel.'

His crew were enraptured, he wrote. 'Never before had Germany's power at sea and in the air been as clear and present for them as it was today. They could hardly get enough of looking at this glorious picture full of strength and greatness.' Meanwhile, he added, 'The Tommies appear to be asleep.'

Adolf Galland, commanding the Luftwaffe escort, could not quite believe that they had got this far unnoticed. 'For two hours – in broad daylight – German warships had been sailing along the English coast along a route no foe had dared take since the seventeenth century. The silence was almost sinister.'

Eventually the British did notice, as the Germans knew they would. But it was too late. A flight of ageing Swordfish biplanes was sent by the RAF to torpedo the warships. Not one returned. Vice Admiral Ciliax and his fleet were past Dover, and Britain had suffered yet another humiliation.

For *The Times*, so often careful in its words, this was too much. There is a time to be a cheerleader, a time to be a critical friend, and a time to excoriate. 'Vice Admiral Ciliax has succeeded where the Duke of Medina Sidonia failed,' the paper of record wrote – naturally assuming that all self-respecting *Times* readers would know the name of the commander of the 1588 Spanish Armada. Agreeing with Galland's earlier assessment, the paper added, 'Nothing more mortifying to the pride of our sea-power has happened since the seventeenth century.'

Churchill could scarcely dispute the assessment. That week he appeared in the House of Commons. There were two items on the agenda: the recent fall of Singapore and what would become known as the 'Channel Dash'. The latter was addressed first: Churchill promised an inquiry. He also promised that its findings would likely remain secret.

In the resulting investigations, there was more than enough blame to go around: the doomed flight of the Swordfish, HMS *Sealion*'s inopportune recharging, the lack of imagination about how bold the Nazis would be.

Of all the findings, the most significant, certainly to Jones, was the one that best explained how they had gone unnoticed for so long: electronic warfare. From across the Channel, devices had been used to block the British radar, transmitting noise on the same frequency. In the skies above, a different kind of deception had been deployed, with bombers picking up the British radar signals and then transmitting them back at high strength – making it look as if one plane was two dozen. The radar was confused, blinded and overwhelmed.

That the Germans had this capability should not have been a surprise – and yet, somehow, it was. Despite using radio counter-measures defensively, to defeat the beams, Britain had been oddly sluggish in developing them offensively. Back in 1939, Cockroft had met with Robert Watson-Watt, inventor of British radar, and, in Cockroft's words, 'discussed with him all the ways we would set to work to fox his baby'. Many of their plans, however, from toy balloons to delayed-pulse oscillators, had in the months since barely progressed from the backs of envelopes.

It was almost as if, Jones suggested, the ease with which enemy radar could be interfered with could not be admitted – because it suggested the same could be done to our own. 'One is tempted to believe that there was an unconscious but deliberate mental avoidance of an unpalatable thought.' Now that it was clear that the same could indeed be done to our own, and just how unpalatable the thought was, efforts were redoubled and nascent schemes were rapidly matured. 'The period of quiescence and consolidation which had elapsed since the decline in activity of the enemy's beam systems was rudely disturbed,' said Jones.

Ultimately this meant that the Channel Dash was, in Jones's opinion, 'perhaps the most important single incident in the history of countermeasures . . . It led to a realistic assessment of the advantages and disadvantages of our own radar systems relative to those of the enemy, and to a realization of the inevitability of a jamming war.' It shattered, he said, 'the palpable fallacy' that we should not jam lest the Germans jammed us. 'The effects on our radar and countermeasures policy were so far-reaching that the escape of these ships, despite the damage to our prestige, was in the long run of the greatest advantage to our overall war effort.'

The RAF's jamming schemes were rushed through. In a jamming war there were, Jones explained, two principles: 'persuading him that you are either (a) where you are not, or (b) not where you are.' The Germans had shown them the tools they needed first. The simplest was a device to block the radar signal – just as had been used by Vice Admiral Ciliax to mask the passage of his flotilla. By broadcasting an overwhelming blast of electromagnetic noise on the same frequency as the radar, it would prevent the Germans from picking out the genuine targets. This would fulfil criterion (b).

The RAF had actually begun, sluggishly, to develop such a system. TRE had shown that six jammers, carried on bombers over the Channel, could reduce the range of early-warning radar to twenty miles. This jamming device, which they called Mandrel, was rushed into service to provide a 'screen' over the Channel. Any reluctance about developing it, for fear that the Germans would then copy it, was overcome by the discovery that the Germans clearly already had it. In this way, the plan was that each night the approach of the bomber stream would be obscured by the Mandrel, until it burst through suddenly, appearing from behind the curtain of jamming.

It worked, in the sense that it blocked radar. But Jones was not hugely impressed. It operated by broadcasting on the same frequency as the radar below. If you took Mandrel on a raid, it wasn't as if the planes disappeared – there was just a different kind of broader signal, a static noise around the bomber force. Over the Channel, that was fine. But since a plane had to carry the device that made that noise and since, over Germany, that plane was normally part of the bomber force, in practice it told the Luftwaffe where roughly to aim. The use of Mandrel screens in the Channel did help – just not to a huge degree. Once the bombers broke through, they were once again in danger. 'The chief tactical advantage gained by the present use of Mandrel is a small degree of nerve strain on defence personnel in France, who cannot relax as freely in the evenings as they could if their early warning system were in perfect working order,' wrote Jones.

He was more impressed with Moonshine, a system that, like the one used by the German bombers, spoofed a radar signal – persuading the enemy you are where you are not. When detected by the Freyas, it would ping back a greatly magnified signal. The first time it was used, in July 1942, 'Three fighters fitted with Moonshine caused the entire enemy fighter force in the Cherbourg area to become airborne,' said Jones.

When the discovery of the German airborne radar Lichtenstein completed the picture of the defensive Kammhuber Line, it was clear that immediate British countermeasures would be needed. It was also clear that what Britain had so far – Mandrel and Moonshine – was not going to be enough. There was one more countermeasure in TRE's arsenal. But, at the Air Ministry, no one could decide whether Britain could risk deploying it.

Countermeasures were not merely a tool for disrupting the enemy's radar. They showed the enemy how to do the same to ours. Early in the war, Cockburn at TRE had suggested jamming the communications of enemy fighters, to prevent them coordinating attacks over Britain. Senior figures in the RAF had fretted that before any such – fairly mild – intervention was made we had to consider whether we ourselves could weather it. Were our radio communications resilient enough? It was a fair point, although also a little like suggesting you shouldn't use artillery until you had a means of stopping an artillery shell yourself.

Even so, this particular countermeasure, codenamed Window, was so effective, so powerful, and, most of all, so simple, that it had the potential, the RAF feared, to devastate radar entirely. If the British used it, would they end up regretting it?

CHAPTER 19

Open the Window

'All warfare is based on deception'

SUN TZU

Buck Ryan's face was ashen. He was Britain's finest secret agent, and he had arrived at a hush-hush facility known as Ack-Ack HQ to convey news of the gravest importance. The Nazis, he had learnt, had devised a plan to blind the RAF at the moment of their invasion of Britain. At a given signal, this secret countermeasure would be activated and their paratroop transporters would sweep in, unseen and unopposed. In the skies above, the brave pilots of the RAF would find themselves scattered and hopelessly confused. This would happen, Ryan knew, because the Germans had found a way to jam radar.

The scheme, he explained to his intelligence handler, was 'simple and based on the element of surprise'. At the chosen moment, Nazi sleeper agents across Britain would deploy box kites, made of light metal alloys. The kites would appear on radar as slow-moving objects. That, said Ryan, was the devilish genius of the plan, 'Everyone's baffled! Orders are on hold until the searchlights sort it out! Now, working to schedule, the Luftwaffe troop carriers fly in to land their Junkers while confusion reigns below!'

With this report Mr Ryan caused considerable alarm in Whitehall. Indeed, it was taken extremely seriously by MI5 – who were responsible for monitoring enemy activity in Britain. When Air Commodore Lywood, head of RAF signals, read of Ryan's discovery he passed it on directly to Frederick Lindemann, scientific advisor to Churchill. This was despite the fact that the place where he had read of the discovery was the comic strip of a tabloid newspaper. Mr Ryan didn't exist. He was the creation of *Daily Mirror* cartoonist Jack Monk.

In recounting the, wholly fictional, secret adventures of the dashing superspy, Monk had inadvertently alighted on a plot device – metal reflectors – that was also a top-secret weapon. It was a countermeasure so classified, so powerful, that the British feared to use it – lest they inspired the Luftwaffe to do the same. Or, as Lywood put it in his memo to Lindemann, 'I don't know whether the Hun is a subscriber to the *Daily Mirror*. If, as I suspect, he is, the attached cutting seems a fairly good investment in basic ideas.'

The principles behind the technology known in Britain at the time as Window, and today better known as chaff, are indeed basic. They were basic enough that a *Mirror* cartoonist could invent and use them for his weekly strip – and that, five years earlier, a young Reginald Jones could casually moot them in a pre-war discussion about radar.

Back in 1937, Jones had been looking at the use of infrared detectors to spot enemy planes – but the upper echelons had begun to cool on the idea. Jones agreed that this form of detection had its weak points, but then so did radar. After all, he pointed out, all you had to do was fly around dropping strips of metal and you could blind any receiver with a dazzling mass of reflections. In 1938 Lindemann even communicated this to Churchill, who warned the enemy could defeat radar by 'strewing' thin wires in the sky.

After the war, some of Jones's contemporaries would criticize him, claiming he took the credit for Window – when, variously, they attributed it to others. It seems entirely possible Jones did invent the idea independently. Equally, it is clear that so did a whole host of others – from a BBC engineer who had, before the war, proposed dragging a wire mesh from an aircraft, to the *Daily Mirror* cartoonist.

This is not surprising: the idea was remarkably simple. Radar works because electromagnetic waves reflect off an object, sending the signal bouncing back. There is nothing unusual about this. All waves, in their way, will bounce off objects. Waves in water will reflect off a harbour wall. Sound will reflect off a tunnel roof – giving you an echo. Sometimes, though, waves do more than merely reflect.

Ask an opera singer to sing in a church and the acoustics will reflect and add richness to her voice. Ask an opera singer to sing a pure enough note, loud enough, and if you have a wine glass of exactly the right shape it will shatter. This happens when the glass is at what is known as the 'resonant frequency' of the note the singer is producing. It vibrates just right, in time to the note – and the vibrations become stronger and stronger, reinforcing with each wave, until the glass can no longer hold its shape.

As with opera singers, so with radar – except you don't need a wine glass. If you have a strip of metal at half the wavelength of the radar then, in theory, it will resonate. When the radar hits it, like the wine glass it will become excited and vibrate in time – re-radiating the signal. So strong is this effect that just a few dozen such strips could look, to the radar operator, like a Lancaster bomber. Drop a few thousand, and suddenly you have an air armada. That, at least, was the theory.

If the idea cannot be attributed to one person, the invention – the turning of an idea into something practical – can. For the Allies, Window went from being an idea to a reality on the kitchen table of Joan Curran, a physicist then in her mid-twenties. By 1942 Curran had already had a good war. She and her husband, Sam Curran, had met while working in Cambridge's Cavendish Laboratory – then the most prestigious physics laboratory in the world. Together, they had spent the early years of the war developing the proximity fuse that would greatly improve the accuracy of anti-aircraft guns.

Then, as part of the scientific team at TRE, she had been tasked with making Window work. So, sitting in her kitchen she began cutting up strips of foil, using the household scissors to methodically snip them to the right length. They were tested using an aircraft sent up from Christchurch aerodrome in Dorset. 'The effects on the radar screens were truly amazing,' recalled Sam years later. 'It looked as if a large fleet of aircraft was present.' Its potential was clear immediately – even if their failure to account for wind meant they had to hastily dispatch police across much of the county to pick up the top-secret tin foil.

There was an obvious objection to the scheme. Fluttering metal strips behave nothing like a bomber – primarily because bombers go forward at 300mph rather than down at 300 feet per minute. But when there are enough of them it doesn't matter. A plane does not move that fast on a radar screen. If the screen shows a range of 100km, it takes over seven seconds for the blip to move 1 per cent of the way towards the centre. So in practice it just takes too long for the radar operator to tease apart the moving signal amid all of the false ones. 'It would probably be possible to demonstrate that any one of these is not an aircraft by fixing it for five seconds or so, but this would be a time consuming process, which would break the heart of most operators,' wrote Jones. No one had the time or the patience to focus on each and every reflection to establish which were actually planes.

Jones calculated that with a steady release of Window, a typical radar station would end up seeing 1,750 spurious responses at any one time. It would be completely overwhelmed – so much so that spotting the reflections moving unusually would be impossible. 'When the false responses are as numerous as the stars in the sky, it will require a certain amount of special training and a certain lapse in time to distinguish which of them are planets, whether the test applied is an object's motion, its twinkling or even its brilliance.' Having met one of those radar operators, after the Bruneval raid, Jones added that he did not see him having the wherewithal to pull off such an analysis. Here, then, was a radar countermeasure that could be made at the kitchen table, and could confound the most sophisticated early-warning devices the Germans could offer. So, you would imagine, the RAF would rush it out? Not exactly.

At TRE a tradition had developed of meetings that they called the 'Sunday Soviets'. In these regular gatherings the hierarchy was flattened. A scientist could argue with an Air Commodore, a private with a Wing Commander. After the Window experiments, a Sunday

Soviet was convened. Of the effectiveness of Window, there was no doubt. 'Discussions showed that the Window method could be used economically to destroy the effectiveness of any radar equipment, and that compared with equivalent electrical methods the effort required was very small,' Cockburn said. But, he added, 'It was also clear that for some time our own radar equipment was probably more susceptible to a Window attack than the enemy's.'

Therein lay the problem. There was a general, probably erroneous belief that Britain needed its radar more than Germany. If the RAF gave the Luftwaffe the idea of Window, the Luftwaffe could soon copy it and seize the advantage. Better, the argument went, not to give them the idea at all, lest the Germans use it to knock out British radar.

Jones, for his part, thought this absurd. He considered it to be based on two completely contradictory beliefs: 1. That the Germans are so clever that they would overcome countermeasures rapidly that would cripple the RAF indefinitely; 2. That the Germans are so stupid they haven't thought of them already anyway. 'It is easy to fall into the trap of thinking that the enemy is omniscient and panoptic and hence of believing that no spoof could fool him. German [radar] personnel are only human, and even a relatively modest spoof might succeed,' he wrote, in response to the first belief. Then, in response to the second, he added, 'It is unwise to be squeamish about taking countermeasures against any enemy development because of the danger of reciprocation. The enemy is not altogether lacking in ingenuity, and has probably thought of most of the counters.'

As it turned out, he was absolutely correct. The scientists of TRE were not, it later transpired, the only ones to be debating the knotty implications of foil. At around the same time as they were holding their Sunday Soviet, a small group of German aircraft were flying over the Baltic dropping bundles of a material they called 'Duppel'. From radar sites across Denmark and northern Germany, controllers watched amazed – just as those in England had – as their screens became a blur of noise. Once again, a single plane was appearing like a massed flotilla.

As the strips of metallic Duppel fluttered into the chilly sea, the presiding officer General Martini prepared a glowing report for

Goering. Goering received it, and was appalled. If used against them, he realized, this simple technology would blast an electromagnetic hole in Germany's defences. The British must never find out. 'We dared not experiment with the little beasts for fear of being discovered,' General Martini later said. 'Had the wind blown when we dropped the metal strips, people would have picked them up, talked about them, and our secret would have been betrayed.'

In a strange twist of game theory, for months both sides held off from the technology in the belief that the other side would benefit from its use more than them.

Finally, after much dithering, Jones's arguments prevailed and the decision was made to deploy Window. Churchill had adjudicated the dispute and with a typical flourish declared his resolution to proceed, saying, 'Open the window.'

The results were more spectacular than they could have dreamed. Later, Jones would say that the high point of the night had been when a German radar operator ordered a packet of metal foil to waggle its wings. 'Impatience rose to exasperation in his voice,' recalled a gleeful Jones. No matter how insistent the operator was, the metal – which he clearly thought was the reflection of one of his own nightfighters – resolutely refused to waggle.

Frank, witnessing the night play out in a listening station, was having an absolutely cracking time. He particularly enjoyed one incredulous announcement, at around 1 a.m. on 25 July 1943. 'Break off!' instructed a Würzburg operator. 'The bombers are multiplying!' Indeed they were.

In the end, several factors had swung the decision to use foil. One was the cavity magnetron. With the ability to make high-powered short-wavelength radio waves, with higher resolving power, the RAF believed it would be better able to cope with the use of Window – or actually, although it didn't know it, Duppel – against it. Another was the incorporation of a new defensive radar into Britain's Chain Home system. This radar operated on a frequency very close to that of the Würzburg – for the Germans, blocking it would also involve blocking their own radar.

The most significant argument, though, was strategic changes in the war itself. Radar remained primarily a defensive tool. The Allies, now, were on the offensive. While earlier the case could have been made that Britain needed its defences more than Germany, now there was no doubt: the Germans were on the back foot, and had far more need of radar.

So on 24 July 790 bombers set off for Hamburg, looking for the tell-tale signature of the docks on their H2S sets. Once in range, they were instructed to drop packets of Window at regular intervals. This was not popular among the crews – who understandably disliked having to keep a door open so they could chuck out a pound of metal every fifteen seconds.

It was even less popular among the Germans on the ground. 'British bombers are too cunning!' exclaimed one radio operator. Another declared it was 'something entirely new!' As the scale of the defeat of their radar warning became apparent, a third said, simply, 'It's bust.'

There had been a plan to print propaganda on the Window strips themselves, making them double, in Jones's words, as 'improving tracts for the German mind', as well as disguising their true intention. Given the many millions that had to be made, this had been viewed as impractical. It was also unlikely to hide their purpose.

It became clear after the war that the Luftwaffe immediately understood what was going on. They just didn't know what to do about it. The first use of Window, said the Luftwaffe general Hubert Weise, was 'shocking'. 'Nightfighting, as well as the aimed and directed fire of the flak, was completely paralysed,' he said.

The RAF's Hamburg mission was codenamed Operation Gomorrah, and it lived up to its name. Hidden from nightfighters, the bombers brought to the city what must have seemed like the wrath of god. The first waves dropped high explosives, shattering windows and doors. Behind them came the incendiaries, catching on the splintered buildings, spreading through houses and warehouses, in a blaze that grew and grew. The world had seen nothing like this. So great was the fire that it became its own weather system, sucking in winds at its base at speeds as high as 150mph. Civilians, fleeing, became stuck in the melted asphalt, screaming and on hands and knees as they burnt to death.

Heinrich Johannson survived by fleeing the air-raid shelter and running with others towards a construction site – where there was little to burn. 'The howling firestorm forced us to use our last strength,' he recalled. Many of those with him failed to make it, and he himself was knocked to the ground by burning debris. Some had taken simply to lying on the ground, protecting their heads. He chose a mound of gravel, burying himself and his wife beneath a blanket that he had soaked in a bath.

'Then I heard a little boy cry, "I don't want to burn! I don't want to burn!" I crawled up to him and brought him back and put him under the blanket with us. Upon my later questions he said, "My mummy is lying dead on those stones, my little brother Manfred lies there too, he is burnt too." His father is at the front in Russia.' For five hours they lay there, watching their city burn, watching people become living torches. Of the 150 in the construction site, fifteen survived.

Henni Flank was one of many forced to make the choice between staying inside and running out of oxygen, and fleeing. She chose, with her baby and husband, to escape into the 'thundering, blazing hell'. The streets were burning, the trees were burning and the tops of them were bent right down to the street.' Horses ran past on fire. They saw someone caught in the wind – a flying, blazing human conflagration.

Above, the bomber crews too knew that something was different. They could feel the heat of the blaze. Some swore they could smell the burning human flesh.

This was a new kind of war. Arthur Harris, commanding Bomber Command, intended to show that bombing alone could break German spirit and win the war. After seeing the destruction Goebbels, the great Nazi propagandist, tended to agree. He called the attack 'a catastrophe that simply staggers the imagination. A city of a million inhabitants has been destroyed in a manner unparalleled in history.' Albert Speer, in charge of German armaments production, warned Hitler that a few more such raids could bring manufacturing to a halt. Over the course of a single week, aerial bombing killed between thirty and forty thousand German civilians, and effectively razed a city.

And each morning, the residents of that city who had survived emerged to see something curious. For three years, an electromagnetic war had waged above their heads, furious but unseen. This battle had filled the air, but it was invisible. Now, it had a physical manifestation. Fluttering down on to the smouldering ruins of the city, settling amid the rubble and embers, was a gentle metallic rain.

For the British, just as important as the German casualty statistics were their own. All this had been achieved with some of the lowest loss rates of the aerial war so far. Of the 790 bombers on the first night, just twelve failed to return.

In the first two months of Window's use, the success – if you could call it that – of the Hamburg raid was never quite repeated. The weather had been unusually dry, the navigation had been uncommonly easy, and the surprise of the countermeasure had since been lost. But, even so, RAF statisticians estimated that two hundred bombers had survived over those weeks that otherwise would have been shot down in flames.

At last, they could show their agents in occupied Europe that their work had not been in vain. One delighted French agent sent a pigeon message, recording an overheard conversation with a German night-fighter control station commander who had declared he would 'rather be attacked by a hundred bombers than submit to that "torrent of paper" and that in the course of one night he had chased 700 bombers without being able to locate one of them.'

Over the weeks that followed, the tactics would be refined. Bomber Command would learn to use Window in conjunction with its other countermeasures. Each night the bomber stream would, as before, burst through the Mandrel screen – suddenly appearing on German radar. But now, so too would fake bomber streams – diversionary Window droppers, pretending to be a massed armada. 'Many German nightfighters would spend fruitless hours and valuable petrol chasing tin-foil in the air,' wrote Jones. And all too often when they realized they were deceived they also found 'a heavy complement of

Mosquitoes ready to attack any German night fighters who might be gulled into an interception'.

It was clear to the Luftwaffe that thanks to the technology they called 'Hamburg Bodies', a complete rethink of strategy was needed. 'The technical success of this action must be designated as complete,' explained one intercepted report. 'By this means the enemy has delivered the long awaited blow against our decimetre radar sets both on land and in the air.'

'Hamburg was wrecked that night,' wrote Jones. 'So was Kammhuber's great, rigid, stereotyped system.' General Kammhuber, creator of the eponymous Kammhuber Line, was moved to new duties in Norway.

But what good would a change of leadership do? There was, for a while, a sense of helplessness. Goering declared his scientists to be 'nincompoops', and then fell into exactly the trap Jones had warned his own superiors against: believing the enemy to be omniscient. 'The British would never have dared use the metal foil here if they had not worked out a hundred per cent what the antidote is,' Goering declared. 'I hate the rogues like the plague, but in one respect I am obliged to doff my cap to them. After this war's over I'm going to buy myself a British radio set, as a token of my regard for their high-frequency work. Then at last I'll have the luxury of owning something that has always worked.'

CHAPTER 20

Feints and Counterfeints: The New Cacophony of War

'I ask you: Do you want total war? If necessary,
do you want a war more total and radical than
anything that we can even imagine today?'

JOSEPH GOEBBELS

August 1943

To the east, the Russians were advancing. To the south, Sicily had been invaded. In Britain, the Allied forces were massing. For the Nazi war machine, things were getting desperate. So no one would have paid much attention to one German nightfighter, downed over Belgium.

The Luftwaffe had been struggling for some time to deal with daytime attacks from the US Air Force, to the extent that nightfighters were being sent up at all hours – even if they were far more vulnerable without the cloak of the dark. In this case, one of the two crew was lucky; he managed to bale out. On the way down, he dropped a small leather bag. As he wafted to the ground, the bag plummeted ahead of him – watched by a member of the Belgian Resistance. The flight documents inside were never reunited with their owner. Instead, the Resistance meticulously photographed each page and, six weeks later, the microfilm landed on Jones's desk.

It was just what he had been looking for. On one of the pages were twenty-one locations, each described in terms of its range and bearing from different German airfields. They were the nightfighter beacons. When the Kammhuber Line had collapsed, rendered useless against tin foil, the Nazis needed a new system – rapidly. Their solution had an air of desperation to it, but also a brute force simplicity. Each night, the radar would attempt to locate the route the RAF bomber stream was taking, and guess the target. Each night, the nightfighters would be directed to wait in the bombers' predicted path – circling around a radio beacon on the route.

As they circled, controllers would provide them with a commentary about the progression of the bombers. And, when the time was right, the same controllers would point them towards their prey and they would surge into the bomber stream itself. The nightfighters' own short-range radar would still give them a hint of the targets and, like lions following the wildebeest, they would pick off the stragglers from the herd. With the stolen map, though, the RAF was able to turn the predator into prey. While the Luftwaffe nightfighters were waiting above the beacon to slip into the RAF bomber stream, the RAF nightfighters could sneak in and mingle among the Luftwaffe.

This was far from the only peril the Luftwaffe faced. Long gone were the days when each countermeasure was debated, the timing of its introduction carefully weighed and argued. 'By now,' Jones wrote, 'all scruples and restrictions had been blasted from the radio war.' As those German nightfighters waited, circling nervously in the darkness, transmitters on the approaching bombers were playing loud noise, broadcast on the Luftwaffe's radio communications frequency, to interfere with the orders that the Germans were receiving from the ground. The closer they got, the stronger the interference. Sometimes they played a noise that the Luftwaffe pilots hinted with some justification was pushing the limits of what was morally acceptable, even in a world war: it sounded like bagpipes.

For Jones, though, the most fun came when the British got their own German-speaking radio controllers involved. From powerful transmitters based in the UK, the RAF began sending their own orders to the Luftwaffe pilots in German. In one exchange, the crowded radio communication bands descended into farce as two controllers both insisted they were the true operator. 'Don't be led astray by the enemy,' warned one German radio broadcast. 'In the name of General Schmid I order all aircraft to Kassel,' he added before – exasperated – he swore. This provided the RAF operator with his opportunity. 'The Englishman is now swearing!' interjected the Englishman in German. 'It is not the Englishman who is swearing, it is me!' came the reply from the German.

In desperation, the Luftwaffe switched to using only female controllers. But there was, as General Schmid – whose eminence had just been invoked – noted, a serious flaw in this counter-countermeasure. 'The use of women's voices proved to be futile, as the enemy had the very same at his disposal.' Or, as Jones put it, 'We had a woman waiting too, and so confusion was immediately restored.'

For a while, the Germans turned to music: a waltz was an instruction to divert to protect Münich, jazz meant Berlin, church songs implied a raid on Münster. Even this code was deciphered.

So it was that the airwaves became a cacophony 'filled with key-tapped long and short waves, which were jammed by whistles, continual dash and frequency sweep', recalled Schmid. There was, he said, 'a mix of reporting orders, talk, music and bell tones', all competing to be heard, or to prevent others being heard. The operator tasked with listening in 'could believe to hear' the noises of a fair during its late evening hours or the transmissions of excited debates in Parliament.

The radio war was now a desperate game of one-upmanship: of measure and countermeasure, of countermeasure and counter-countermeasure, of counter-countermeasure and counter-counter-countermeasure. As Schmid put it, 'The development of radar on both sides during the war may be compared with the artillery of a ship and the armour plating in the battle of navies.' In other words, an arms race was rapidly developing, in which better weapons competed with better countermeasures. After the Luftwaffe equipped nightfighters with onboard radar as standard, the RAF designed Serrate – a device in its own nightfighters that locked on to the radar signal. After the RAF designed Monica, a system that detected Luftwaffe radar to warn of impending attack, the Luftwaffe designed its own system that homed in on Monica – from far further away than airborne radar could have achieved. 'The very device which was intended to enable our bombers to evade the German fighters had in fact the opposite effect,' lamented Jones.

Like the undersea soundscape, in which each click and whistle, each burst of sonar, each fish evading the sonar, is part of a furious evolutionary competition, there was a great profusion of countermeasure species. The Allies had Grocer and Airborne Grocer, Cigar, Jostle and Piperack, Bagful, Blonde and Boozer. The Germans had Barbara and Barbarossa, Bernhard and Bernhardine, Taunus and Berlin. Each fought for supremacy in its own band of the electromagnetic spectrum. If each bought as much as a six-week advantage before it was countered, it was enough.

These were desperate months, for the scientists and especially for the crews. The problem was that all too often it was the case that each new technology, by the very nature of how it worked, was the blueprint for its own countermeasure – by broadcasting on a set frequency it inherently contained the technical specifications that facilitated its own undoing.

Sometimes, though, to Jones's great frustration, sheer carelessness meant the Germans had more assistance in confounding the British efforts than they should have. One of the best examples was the way the RAF used H2S – the airborne radar that provided a map of the ground below, and which was being rolled out as standard across the bomber fleet. It was barely days after that first successful raid, when the shape of the German coast had been so clear, that the first H2S-equipped bomber was shot down. This in itself was inevitable – and expected.

Having carefully reconstructed the shattered equipment, the German scientist Professor Leo Brandt took it up to the roof of a tall building in Berlin. From there he could look down on the city laid out below him – seeing 'excellent images of the Berlin residential area'. He noted that the lakes, in particular, stood out. Now, just as the British had used fires to deceive Luftwaffe bombers about the location of towns, the Germans devised decoys to trick H2S. In Kiel harbour, a replica of the port was designed and floated, using specially designed 'corner' reflector buoys – containing sharp angles that, in theory, sent back a strong signal.

Far more significantly, though, the Luftwaffe also designed a device known as Naxos that could lock on to the H2S signal. It was this that led General Schmid to declare the use of H2S the 'greatest single mistake' in Allied bomber policy. The mistake was not really its use, though. It was its blanket use. And this mistake was doubly inexcusable because of what had happened a few months earlier, with an entirely different radio aid.

The skies above the south coast of Britain were a busy place. There were incoming raids and outgoing sorties. There were damaged bombers limping back from the Ruhr, and lone Luftwaffe flights probing defences. All were plotted on Chain Home, Britain's defensive radar system. But how to tell one from the other? The answer was IFF – Interrogate Friend or Foe. This was a system for making sure that friendly aircraft were clearly identified. When a plane carrying an IFF device was pinged by the Chain Home system, it re-radiated the signal, leading to a stronger blip – telling the radar operators it was a friendly plane.

This system was only needed while planes were in range of Chain Home. But RAF bombers found another use for it. When caught in the enemy's radar-guided searchlights, they found that if they switched on IFF they could escape. No longer were the searchlights able to lock on so effectively. Instead, they were – the pilots came to believe – confused by the IFF signal. So they left the sets switched on. When Jones first heard about this, he was incandescent.

Bomber crews were a superstitious bunch, and believed many things – of course they did. When, each night, you put your hands in fate, taking a one in twenty chance of being shot down, knowing that, eventually, probably through no mistake of your own, you would become one of the one in twenty, you took comfort where you could. In a life ruled by randomness, what else is left but superstition? You followed a 'lucky' pre-flight routine. You avoided the barmaid known as 'chop blonde', whose lovers always ended up dead. You carried a photograph of your loved ones. James Insull, a navigator, claimed he was 'flak-proof' – 'but I do always take my pair of dice and my daughter's wee bootie with me on every trip'. It was not obvious whether this was despite his being flak-proof, or what made him so.

For Jones, it was clear that the IFF searchlight interference was just the radio equivalent of the wee bootie and lucky photograph. Almost certainly, he explained patiently, it didn't work. On the off chance it did work, though, he added, it was catastrophic. Here was a signal that the Germans could already spot, and if they could spot it, they could track it. So worried was he that he gave talks to airmen, warning of the danger. To his amazement, they argued back. IFF did confuse searchlights, they insisted. They were, we now know, completely wrong, and Jones was completely right. It was just another superstition – with the pilots attributing chance to design.

He was also right that the Germans had spotted it. Using a device that mimicked the Chain Home radar, the Germans had found a way to challenge IFF and lock on. Then any bomber stream in which a plane had its IFF switched on could be tracked. Even if most bombers followed the rules, it didn't matter. All it took was a few superstitious airmen among the bomber stream to give the game away. 'If the Germans had found one bomber, they had probably found the lot,' lamented Jones. 'Our whole stream was thus betrayed, and its intentions divined, so that the German nightfighters were not only able to identify and to reach the target earlier, but also to mix in with the stream on its outward and homewards journeys.' It took concerted efforts at education and countless losses to enforce IFF silence.

All of which should have been a warning: that any radio device in the future should only be used when it was needed. If it was, it was not a warning heeded by the bomber crews. Excited by their flashy new H2S apparatus, they took to switching it on to warm it up while still on the runway. And, as they rose above that runway, the Luftwaffe spotted them.

It was, said Schmid, 'a small British carelessness'. Window had destroyed the Luftwaffe radar network, but it turned out that thanks to H2S they did not need it. 'The German detecting service was able to follow closely the path of the British bomber stream from the rally point to the target and back to the landing,' said Schmid. As Professor Brandt put it, for a few vital months, 'from the moment they took off until the moment they landed again, the pathfinders leading the squadrons were completely under surveillance.'

By now, Jones was the head of a proper department. Still a small one, but perfectly formed. In 1941 he had gained three staff, including Mary Francis, a mathematician, and another 'girl', Ginger Parry. In 1942 there were two more – both men, one a scientist. In 1943 there were four more, three of them scientists and the fourth a military specialist: Wing Commander Rupert Gascoigne-Cecil.

Now once again, Jones and his growing team sought to demonstrate that this simple and easily remedied mistake had undone much of the radio countermeasures work. It took, he said, 'an enormous amount of proof' before they would be believed. 'When we did finally get them all off, the Germans couldn't believe it: they thought we had changed our frequencies, they spent nights searching the ether for new frequencies . . . but it was an awful struggle.'

So it was that out of the cacophony, the great profusion of noise that occurred in this first flourishing of electromagnetic conflict, came a measure of silence. 'The enemy and ourselves had mustered considerable homing effort,' noted one report. 'The previous extravagant use of radar was no longer possible without serious disadvantage to the user.'

'Bomber Command,' wrote Robert Cockburn, from TRE, 'had to weigh up very carefully the advantages to be obtained from radiating devices.' Each new device, now, was part of a careful calculus in which it was never immediately clear that the benefits from using it would exceed the vulnerabilities it introduced. Slowly, though, the electromagnetic battle turned. By 1944, it was going so badly for the Germans that the Luftwaffe offered an essay prize, asking radio operators for their thoughts on tracking Allied planes. (Jones, having received the intercepted message announcing the competition, was very tempted. 'Only with difficulty could we restrain ourselves from entering . . . With luck we should have won.')

It is tempting to look at all the brainpower on offer and conclude that victory in the radio war would go to the brainiest side. It is true that the Allies were not short of ingenuity – but neither were the Germans. Until the end, they were still coming up with new ideas, still planning the defensive move that would counter the Allied attacks.

There was a scheme to install teleprinters on planes, so that plain-text orders would come through – it is far easier for a machine to ignore interference than a human. There was a fiendishly clever project to use the Allies' own radar against them – the deflections of the Chain Home system, bouncing off the side of Allied bombers, could theoretically be picked up and triangulated in France, providing a way to peer through the Mandrel screen. There was a modification made to the Würzburgs that could 'see' through the Window, spotting the slight modulation in the signal made by an aircraft propeller and adding an aural humming sound to the visual echo. It was ingenious, but also desperate. 'In practice,' wrote Jones, straining to hear the signal 'seriously decreases the range . . . and impairs the sanity of the operators.'

The radio battle, ultimately, was lost not because the Germans were less clever, but because they lost the ability to wage it. 'The German players lay on the ground,' said Schmid after the war, before adding, with an endearing attempt at idiom, 'They could not use their balls anymore.' Radio operators were drafted on to the front-lines, planes grounded through lack of fuel. They could no longer apply the resources to bring their brain power to fruition. 'All technical progress was of no use anymore – they had come too late.' But, noted Cockburn with evident respect, on occasion it was a close-run thing. 'Although the ultimate issue of the radio war was for a long time in doubt,' he wrote after the war, 'in the final outcome the enemy's defences were completely neutralized.'

That final neutralization of the Nazi radar defences began with a great battle, the greatest battle of the war so far – and, although the part played by radar is rarely told, it was also the apotheosis of the Allied radio offensive: D-Day.

CHAPTER 21

D-Day

'In wartime, truth is so precious that she should always
be attended by a bodyguard of lies'

WINSTON CHURCHILL

June 1944, the English Channel

ALONG THE ATLANTIC COAST, from northern Norway to northern Spain, was a great profusion of radar. New, old, experimental, obsolete – all jostled to saturate the airwaves. Between Le Touquet and Le Havre, there was one radar station every three miles. It was as if the Luftwaffe hoped to overcome countermeasures with sheer quantity. But, said Jones, the advantage gained, particularly in a time of Window, 'bears little relation to the effort involved'.

In a series of documents, he and his colleagues methodically listed each of the radar towers for which they had locations. Some had been identified by the Resistance, some by Enigma intercepts. Some had been seen from above, by the increasingly sophisticated photographic interpreters – squinting to see the fuzzy shadow of a moving aerial. Some had been tracked down using 'ping pong', a specially designed radio triangulation device – which was itself only possible because of the carefully accumulated intelligence on which frequencies to listen out for. Later, they got to check their efforts against the Luftwaffe documents. 'Between Calais and Guernsey there were 120 pieces of equipment . . . we found them all.'

There was, of course, a particular reason why they might be interested in that particular stretch of coastline at that particular time. In the introduction to one of the lists, Jones stated they were writing it 'anticipating the reopening of the First Front'. D-Day was coming. Never before in the field of human conflict had so big a secret been known by so many, and not managed to leak. It was the job of Jones and his team to help protect that secret even through the very last hours. They needed to prevent the invasion fleet being spotted as it crossed the Channel. To do this, in the weeks preceding the invasion Jones had liaised with supreme headquarters about undermining the German early-warning system.

It was a delicate task. The RAF had to obliterate the chances of the Germans spotting the fleet attacking Normandy without giving a hint that Normandy was the landing ground. They also had to leave just enough radar in place so that that which remained could be deceived. In the end, they came up with a simple rubric. 'It was arranged to attack two stations outside the landing area for every one within it,' said Jones. Using fighters armed with rockets, the Allies launched 1,668 sorties – flying low-level raids at an eye-watering cost in pilots. One sortie, on Cap de la Hague, affected Jones greatly. After the invasion he had received an eyewitness report from a German. Three pilots had approached in line astern. One was hit by flak and dived his doomed plane into the radar, destroying it. 'The German said it was the bravest thing that he had ever seen.' Jones wanted to get him a posthumous Victoria Cross, but couldn't. Of the two surviving planes one had been shot down on the return, and it couldn't be established which had destroyed the radar.

By the night of 5 June just a small handful of radars remained. Enough had been destroyed that they missed the launch of the largest invasion fleet in history. Enough survived, though, just where they needed to be to witness the launch of what had to be the strangest.

While seven thousand vessels quietly left dock headed for

Normandy, over to their east a handful of small boats lumbered into the Channel, some heading for Pas de Calais, some for Cap d'Antifer. Some towed a massive barrage balloon behind them; some carried on deck the radio device Moonshine. Above them, just visible in the dull cloudy night sky, little strips of foil fluttered to the water. Here, for the second time in the war, and exactly four years after Dunkirk, a flotilla of little ships, apparently ill-suited to war, set forth for France – this time in an offensive role, of sorts.

Slowly, this fleet headed towards the French coast. In the sky, working in shifts, bombers flew in a tight oval over the boats. These planes, which included the famous 617 'Dambusters' squadron, had been training for weeks. They had been learning to fly circuits that moved forward at the pace of this dummy invasion force – using the new navigational device Gee, which painted invisible grid lines in the sky, to advance exactly 0.82 nautical miles each time round. As they flew, they dropped a curtain of Window into the sea, thousands of lengths of metal slowly dropping to the waves. As the late evening midsummer dusk gave way to the darkest part of the night, on board the ships the Moonshine equipment at last registered some inter-est. Moonshine was designed to spot radar signals and return them – but do so amplified.

On the distant coast, which was by now murkily appearing in the gloom, the Moonshine on the ships told the fleet that the radar of patrolling German aircraft had seen them. This was as they had hoped. By this stage, all but one of the Moonshine operators was suffering from seasickness, but still they sent back the signal on the same frequency, strong and unambiguous. It was, it would have appeared, the signal of a large fleet.

Above, just so the message could not be misunderstood, the barrage balloons offered a reflecting surface equivalent to a 10,000-tonne ship. Luckily the coastal Freya systems were not sophisticated enough to spot that this large craft was apparently levitating.

The point of the fleet was to be seen, but to not look like it wanted to be seen. The Navy provided just enough jamming to add plausibility, without actually being effective. For the remaining German radar operators on that June day it was clear what was going on: the long-expected invasion was coming. And, it seemed, it would be landing in the Calais region.

Here, in this bizarre collection of boats, balloons and planes, was the logical conclusion of the war of the airwaves. Had any German seen the fleet leave, it would have inspired mirth more than fear. But it was not designed to be seen – or, at least, not seen using visible light. This was a fleet for the electromagnetic war and, as a collection of bright dots on a cathode-ray tube, it was as scary and formidable as any the world had witnessed.

What, though, if a plane had been dispatched to gain visual confirmation? As it happened, none was. Cockburn, whose TRE scientists helped devise much of the scheme, suspected it wouldn't have mattered. Their feint relied on radar science, but also psychology – and an understanding of how large organizations worked.

After the war Alfred Price, the great radar chronicler, interviewed Cockburn. 'He said one should try to imagine the scene: a frightened under-trained young conscript radar operator sees the "ghost" fleet on his screen and reports it to his headquarters as the long-expected enemy invasion force; so do his colleagues at radars on either side. Soon their plot is shown as a nice broad arrow on the situation map at headquarters. The "ghost" fleet is now a military fact. If aircraft were then to fly into the area and report it clear of ships, would their reports be believed? Probably not. The operation was to take place at night and the crews might not be where they thought they were. Once a broad arrow representing an enemy attack appeared on the situation map at headquarters, Cockburn believed, it would take a lot to remove it.'

By the time it stopped a few miles off the coast, the fleet had become that broad arrow – a fact on a map and thus a fact on the ground. As, a few miles along the Channel, a real fleet carrying real soldiers readied itself for the battle ahead, here the small crew switched on speakers that played the sounds of a great armada weighing anchor. Their work was done.

The deception, though, was not. The ghost fleet was just phase one. No modern invasion is complete without its air support.

*

No one would ever claim that Rupert represented the *crème de la crème* of the British Parachute Regiment. For one thing, he was rather short. For another he was dropped with little in the way of firepower. Primarily, though, the reason that he would have done little to scare the Wehrmacht is he was made of stuffed hessian. Yet, despite everything, the Wehrmacht were indeed scared – for a few hours at least. And a few hours were all that was needed.

If the real D-Day invasion fleet was vulnerable to attack, the Allied high command was acutely aware that the men due to land behind enemy lines, flown in to hold key strategic targets, were even more so. 'Two large formations of slow, unarmed aircraft had to fly in moonlight across an area in which the enemy could deploy his night fighters,' wrote Jones. 'It was clearly necessary to prevent him from making a successful interception of these forces.' This was why at the same time as the ghost fleet was preparing its ghost landing craft, ghost flights were setting out from the airfields of Kent.

Each had a different role. Some were there to simulate bomber raids, drawing fighters away from the main attack. One group was tasked with dividing France in two, flying backwards and forwards along the line of the River Somme. This small group of brave fliers – each aware they were bait – dropped bundles of Window and blasted out jamming signals. Their job was to create a screen, electromagnetically cutting France in two.

Jack Furner, a young navigator who went on to be an old Air Vice-Marshal, remembered the excitement of being on that flight, creating 'an electronic wall' that shielded the '1,000 Allied transport aircraft that were on their way to drop paratroops' to the south-west.

To the north of that wall came the Ruperts – an airborne force of marginally more sophisticated scarecrows, stuffed hessian sacks with parachutes attached. Almost four hundred would make the ultimate sacrifice for King and Country that night, sometimes in dramatic fashion. Dropped into the fields in northern France, some contained timed explosives – designed to simulate a burning soldier in his parachute, or to create the sounds of battle on landing. Alongside them were a few members of the Special Air Service. Their orders? To make a lot of noise.

It worked. They created just enough confusion, for just long enough. That night, far from the safety of the real invasion force and with only stuffed Ruperts for back-up, these soldiers landed in the forests and fields. On maps in local headquarters more arrows appeared, depicting phantom thrusts. At a time when the Nazis should have been diverting troops to Normandy, they were chasing shadows – and hessian – far from the real battlefield.

'Your task will not be an easy one,' said Dwight D. Eisenhower, in his eve-of-battle message. 'Your enemy is well trained, well equipped, and battle-hardened. He will fight savagely.

'But this is the year 1944. Much has happened since the Nazi triumphs of 1940–41. The United Nations have inflicted upon the Germans great defeats, in open battle, man-to-man. Our air offensive has seriously reduced their strength in the air and their capacity to wage war on the ground. Our Home Fronts have given us an overwhelming superiority in weapons and munitions of war, and placed at our disposal great reserves of trained fighting men. The tide has turned. The free men of the world are marching together to victory.'

D-Day, Eisenhower's Great Crusade, succeeded because the Allied ships had supremacy in the sea. It held its beachhead because the planes had supremacy in the air. What he could not say is that for the first time in the history of warfare, the Allies had another kind of superiority too: superiority not just of the air, but of the airwaves. In the last year of the war, the electronic victory predicted the physical one. On some nights, wrote Jones, 'all the forces would be spoof ones.'

The Luftwaffe situation was increasingly tragicomic. 'With their radar rendered useless, their radiotelephony issuing contradictory orders, and their petrol being wasted, the troubles of the German nightfighters were not exhausted. Whenever they took to the air they were a prey to our-long range nightfighters, now hunting in force over the Reich,' he wrote.

When they returned home, meanwhile, the airfields would be blacked out for protection – meaning many fighters crashed on landing. For those that did make it back, 'As a final pinprick, General

Schmid would castigate the night's survivors with admonitions and threats.' In a separate report, Jones summed up their situation more pithily. 'Frustrated, tormented and thrashed, the German nightfighters never recovered.'

The eventual defeat of the Third Reich's electromagnetic defences was the work of many. It involved pilots, commandos and resistance fighters – brave men and women who fought and died, sometimes to send back the details of technology they themselves did not understand. It involved the vast network of spotters, listeners and technicians. It involved the engineers, the tinkering 'backroom boys' (and women) who turned equations into circuits, valves and aerials. It involved a small group of scientists with the particular skill of putting themselves in the minds of their Nazi counterparts.

And, within this group, it involved one man in particular – a man with the vision to see the vast possibilities contained in the light beyond the rainbow.

Epilogue

'Noble efforts in the high air and the flaming streets would
have been in vain if British science and British brains had not
played the ever-memorable and decisive part'

WINSTON CHURCHILL, *THEIR FINEST HOUR*

15 February 1945

IN THE LAST WINTER of the war, a memo arrived in Reginald
Jones's in-tray. His department was still busy. It was receiving
intelligence on the V2 rocket programme, reports on the death throes
of Berlin's radar defences, and speculation about the location of
nuclear scientists. Now there was yet another document vying for his
attention. It was titled, 'Problems arising out of the termination of
the war with Germany'.

Bureaucracies gain their own momentum. Even in wartime there
are jobs that, once created, need to be performed; roles that create
their own role. So it was that someone, somewhere, was tasked with
thinking through the grim possibility of national rejoicing. And they
were very worried.

'Higher authority is of the opinion that the exuberance shown by
service personnel and civilians ... may cause damage to Government

property', they explained, adding, darkly, that their shrewd understanding of human nature suggested 'festivities may be prolonged'.

While the possibility that government assets might be harmed was uppermost in the author's mind, a secondary consideration was the safety and virtue of the Women's Auxiliary Air Force, and the 'dangers of being on the streets during the convivial period'. To this end, the document contained detailed plans for saving those women from the perils of partying, educating them on avoiding the streets, and escorting them home.

Tickled by this, Jones decided to have some fun and drafted his own memo in response. There were now several women staff working for him, he stated, some in technical roles. One, Flight Officer Margaret Masterman, was a particular concern.

'I think you should know', he wrote, 'that Flight Officer Masterman has been showing repeated signs of conviviality at the slightest excuse. On Christmas Day, for example, she allowed herself to be kissed under the mistletoe in a tavern, announcing afterwards to one of my officers that on such occasions "one mustn't be cagey".'

It got worse. 'On the inauguration of the New Year she went to a party; at midnight, after singing "Auld Lang Syne" she kissed her neighbours more than once, with the remark, "In my part of the country we always kiss twice".'

He fretted that her 'wantonness presents special difficulties' and asked that she be signed up immediately for any classes on street-avoidance. After all, he said, if she were this 'uncagey' for 'such trivial festivals', 'you will understand my very serious misgivings regarding the limits to which she might go on the outbreak of Peace.'

Yet peace did indeed break out. Masterman, and her colleagues, having survived the conviviality of VE Day, began looking for other jobs. Britain was broke. The vast wartime infrastructure needed to be disbanded, and although Jones would stay in his job into 1946, and could have stayed longer had he wanted, he too felt it was time to leave.

But where to go? In the spring of 1946, he travelled to Aberdeen to be interviewed for a chair in physics at the university. He had good reason not to get his hopes up. Two of his competitors were already established professors at universities. He, in contrast, had left academia almost a decade earlier, as a junior researcher. In the years that followed he had published nothing. The qualifications he did have were considerable, but classified.

So he was surprised to be offered the position. He asked the Principal of Aberdeen what had swung it. 'A friend of yours was here last week,' replied Sir William Hamilton Fyfe. Who was it? 'Winston Churchill.' Churchill had been up to receive an honorary degree, and Fyfe complained that every time they met he badgered him about the appointment. Churchill was adamant: he simply must employ the man 'Who Broke the Bloody Beam'.

Postscript

Here is the story, as I heard it.

It was early June, 1944. On the south coast of England Mary Moore, a radio scientist, was working her way between ships, methodically briefing the navigators. Ahead of the Normandy invasion, select vessels, among them the minesweepers that would clear the way, had been given a new piece of radio equipment that would help them to track their position as they crossed the channel.

Patiently, Moore, who had been sent by the Admiralty, explained the system – how the navigator would twiddle the dial beneath a screen to bring the two sinusoidal curves together, each representing a timing signal from different transmitters. At the back of her mind, though, was her own timing signal. She did not know when, but, at some point, to use the phrase of the day, the 'balloon would go up'. D-Day would be on. Then no one – whether sailors, soldiers or mathematicians – would be able to leave the ships. If it happened while she was onboard, she would be on her way to Normandy.

Apparently, she got off just in time.

Here is another story. A few miles down the coast at Portland, a few hours earlier, Raymond Whipple, another scientist employed by the Admiralty, was also inspecting the fleet. Puffing on his pipe, he was the picture of an absent-minded professor. He, too, had been involved in developing radio systems.

Earlier in the war, in Portland Harbour, he had been given a captured German radio-controlled torpedo to play with. He had had a lovely time – until he lost it. He was having a lovely time that June day too. Although he had a view of the greatest invasion force in history, he had been distracted by a particularly interesting cloud formation. To the delight of the sailors, he walked right off the pier into the sea.

History is about all of us. It is about humanity in the aggregate. What we humans like, though, is stories in the singular. Even when it comes to the Second World War, the greatest struggle humanity has waged, we like to understand the narrative through tales of individuals.

I could have tried to tell the story of Mary and Raymond. But for them, the anecdotes above are almost all I know of their war work. Like most of those in radio research, they left no diaries, and rarely spoke of that time. Their research was important – I have no doubt about that. According to one half-remembered story, told by a contemporary, Raymond solved a mathematical problem that enabled the more rapid detection of submarines – shortening the war, the colleague estimated, by a week or more.

But many people's research was important. Many people helped find submarines.

Unlike with Jones, with Mary and Raymond all I have are these anecdotes that I'm not even sure are true. They have been told and retold. They are filtered through time and through telling: they are great fun, almost certainly because their inconvenient edges are lost in the service of their audience.

Are they the truth? They are the closest I can get to it. Would I be surprised if the details, perhaps important ones, are wrong? I would not. And yet, few people in the world have better claim to knowing their story than me.

Mary and Raymond, you see, are my grandparents.

I never knew Raymond, he died before I was born. I was always conscious that at university I followed in his footsteps, studying mathematics at Cambridge. I was always conscious he was a lot better at it than me. I did know Mary – Granny Mary, we called her – a bit better. She died when I was ten. Even so, all I have of her are snap-shots. In my mind I have joined these into a narrative of sorts, patched together with the fallible glue of memory.

I remember her taking me to the pantry, where she told me about sandwiches with dripping in. I remember her visiting when we dug the pond in our garden. I remember her sitting in a sunny garden, smiling – or do I just remember a photograph of her sitting in a sunny garden, smiling? I have a memory, a feeling might be a better word, of a woman in sturdy, no-nonsense hiking boots, cooking sturdy, no-nonsense meals for a large and growing flock of grand-children. She did the things grannies should, but always with a slight hint (or have I imagined this too?) that she could do more.

Nowadays, I look back on her a little differently from how I saw her then. I try to imagine her as a human on her own, rather than merely someone defined in relation to her children and grandchil-dren. She was a housewife, but also a key member of what these days would be called a town's 'civic society'. She was a magistrate. She vol-unteered for things. In her home town of Wantage, there is a road named after her.

It's hard, as I said, not to make a narrative when gluing together these memories. So here is my analysis of her. During the war she was in a job of importance, respected for her intellect. After the war, when Raymond went into nuclear research, she went into the work expected of her: having a family. It feels, to me, like she spent the fifty years afterwards trying and failing to replicate that excitement and intellectual vigour of those war years. But – and this is the dilemma of the biographer – what do I know? How can a few hundred words – or a few tens of thousands – ever do justice to a person? How can it hope to even approximate to truth?

This book is my best stab at telling the story of the radio war. Because we are humans, and we like stories about people, I tried to tell it through a person. The truth is, when picking which of those individuals to follow, we tell the stories we can. Jones kept many papers. He wrote much and had much written about him. I can tell his story. Jones more than deserves to have his story told, but his is just one story. Raymond and Mary had their own story, and for me they represent all the thousands of radio researchers whose stories I haven't told, whose stories no one will ever tell.

I dedicate this book to my grandparents. And I also thank them for what they did, personally and professionally, in the war.

Mary and Raymond met researching radio waves. They met because humans were hubristic enough to think they could harness, and manipulate, the light we cannot see. Because they met, I exist.

Notes

My book traces its genesis to a gloomy high table at Queens' College, Cambridge. I was visiting Ramsey Faragher, an old university friend who had a position teaching there. He had also just started a company that was based on commercializing his PhD – looking at improving the GPS positioning used by mobile phones in built-up areas, where the radio signal bounces and reflects off buildings. Over the course of dinner, he told me about this book he had read, an autobiography of a radio scientist who was a legend in his field. In the book he described how he had worked out that the Germans were painting electromagnetic crosses over their targets, and that he had helped defeat them. Ramsey suggested that I should write about him. I was already writing another book but, on the train home, I ordered the autobiography. And, two years later, I got around to reading it.

That book, *Most Secret War*, by R. V. Jones, is rightly legendary. I am neither trying to replicate it nor replace it. My book is doing something different. It is contextualizing it, broadening it and also – sometimes – narrowing it. Inevitably, especially when dealing with personal anecdotes, his autobiography provides a decent chunk of my source material. The war work of Jones is, though, merely the spine I use to tell the tale of the radio war. On that spine I hang the tales of many others – the engineers, commandos, agents and Germans who also fought.

I use documents from the National Archives and from Jones's own papers. Once top secret, now declassified, these documents go far beyond those simply relating to Jones's work. It felt absurd that some-one like me could turn up, get a reader's card, and hold something once read by Winston Churchill. More than that, something that once contained information that ruined Churchill's day.

I make no pretence that I am the first to write about the radio war, but I think I am the first in at least forty years to tell its tale in one place. The time that has passed allows perspective; it allows more information – from all sides – to offer context, and temper myths.

One source, also listed in the bibliography, is the PhD thesis of James Goodchild, a historian. It is impossible to immerse yourself in the world of Jones and not accept he was a showman who sometimes enjoyed a good tale a bit too much. Goodchild's thesis, 'R. V. Jones and the Birth of Scientific Intelligence', is a nice foil to some of his excesses. Goodchild sought to counter what he saw as an overreliance on Jones as source material in historical accounts of the time. His research was not only, for me, a summation of useful context, it was also something of a relief – to know that someone had gone through the archives with a professionally sceptical eye towards Jones's account, and found that, with some of the showier edges knocked off, it broadly held up.

My book does not pretend to provide a definitive historical account. More than anything, I want the book to be read. I have left out much of the radio war – because covering everything greatly increases the chances readers give up, and end up discovering less. In the bibliography I have listed books that can offer more depth for those who are interested.

I have left out much of Jones's other work – his battle against the V2 rockets is the subject for another book. I have skated over some of the trickier technical problems – I view my readers as intelligent non-specialists, and part of respecting them involves knowing when you are going too far into the weeds. This is, most of all, a book I want people to read and enjoy – a racy story of clever people doing important things, packed with anecdotes and tales that I hope people will want to pass on.

Two archives in particular have been invaluable: the National

Archives, whose references I will prefix TNA, and Churchill College Archives, whose references I will prefix CA.

I am extremely grateful to the Medmenham Collection for their kind assistance.

Prologue

1 **Reginald Jones was late:** Jones's climactic meeting with Churchill is recounted by him, in *Most Secret War*, and corroborated by Churchill himself in *Their Finest Hour* (volume 2 of his *The Second World War*), which is where the Churchill quotes in this chapter come from. The meeting is also referenced in passing by Harry Spencer, in CA ROCO 1/1

5 **'on a typical operation involving 750 bombers, it meant the loss of 260 aircrew':** Furner, Jack, '100 Group. "Confound and . . .", *Royal Air Force Historical Society Journal*, 28, 2003

5 **only one in three actually got within five miles of it:** *Butt Report*, 1941

PART 1: DEFENCE

1. *The Killing of a Sheep at a Hundred Yards*

13 **Would it be possible, he asked, to borrow a typewriter?:** The account of the circumstances around Hans Ferdinand Mayer's intelligence leak comes from a patchwork of sources, some contradictory. For instance, exactly what was said on the BBC World Service, and when, is far from clear. I have patched together these sources as best I can. The bare bones of what was received and when come from the Oslo Report itself, and British intelligence reports – among them CA RVJO B.24, Air Scientific Intelligence Report No. 13 'D.T', 10 January 1942. The best accounts of Mayer come from *Reflections on Intelligence*, a 1989 book by Reginald Jones, pp. 265, 284, 333, 320–327. I have also included information from *Anpassung, Unbotmäßigkeit und Widerstand*, by Joachim Hagenauer and Martin Pabst, from the Bayerische Akademie der Wissenschaften. The title translates as 'Adjustment, insubordination and resistance'.

17 **'I have seen my death':** Macfarlane, Ross, *X-Ray Anniversary*, Wellcome Blog, November 2010, blog.wellcomelibrary.org/2010/11/x-ray-anniversary/

17 **Percy Spencer was working on microwave emitters:** Ackerman, Evan, 'A brief history of the microwave oven', IEEE, 30 September 2016

18 **'The New Death-Dealing "Diabolic Rays"':** *Popular Radio*, August 1924

19 **'demonstrate the killing of a sheep at a range of 100 yards':** Rowe, Albert, *One Story of Radar*, p. 6

19 **'Attention is being turned to the still difficult':** Allison, David Kite, *New Eye for the Navy*, p. 142

20 **'I think it is well also for the man in the street':** Hansard, 10 November 1932

20 **'more important than a cure for cancer':** Judkins, Phil, 'Making Vision into Power', PhD thesis, 23 May 2008, p. 290, https://dspace.lib.cranfield.ac.uk/handle/1826/2577

20 **'decorous atmosphere':** ibid, p. 184

21 **'Silhouette':** ibid, p. 278

21 **a Heyford bomber flew:** ibid, p. 108

22 **'Britain has become an island once more'** . . . **'gravely interfere with grouse shooting':** Clark, Ronald, *The Rise of the Boffins*, pp. 37–41

23 **'I don't like boxes with coils':** Price, Alfred, *Instruments of Darkness*, p. 48

24 **Philipp Lenard:** Nobel Prize biography, https://www.nobelprize.org/prizes/physics/1905/lenard/biographical/

25 **'planted by a wily enemy':** CA RVJO B.24

2. Reginald

26 **As Harold Jones and his comrades marched past:** Jones, *Most Secret War*, p. 12

27 **The family lived in Dulwich,** and childhood accounts: ibid, pp. 3–7

28 **The most exalted prize on such a night was a policeman's helmet:** Jones, *Sir Charles Frank, OBE, FRS: An Eightieth Birthday Tribute,* CRC Press, 1991, p. 4

30 **'Love me, love my dog' and account of road trip:** Mukerjee, Madhusree, 'The Most Powerful Scientist Ever: Winston Churchill's Personal Technocrat', *Scientific American,* 6 August 2010

30 **'A terrible process is astir':** Churchill, Winston, 'How I Would Procure Peace', *Daily Mail,* 9 July 1934

31 **working in Oxford's Clarendon laboratory:** Jones, *Most Secret War,* p. 21

32 **'dropping a brick',** and Thost's story from the MI5 end: TNA KV 2/952

33 **'A newspaper correspondent might easily be a cover-operation for a spy',** and Thost's story from Jones's end: Jones, *Most Secret War,* p. 27

34 **He was still at the Clarendon in Oxford, but now the government was paying his wage,** and Jones's career progression: ibid, pp. 21–50

35–6 **'My search through the files taught me how primitive':** From a lecture given by Jones to the CIA, 1947

3. *Reading the Tea Leaves*

37 **There was to be a 'symphony of death',** and the story of De Wohl: Winter, P. R. J., 'Libra rising: Hitler, astrology and British intelligence, 1940–43', *Intelligence and National Security,* 2006. 21:3, 394–415, DOI: 10.1080/02684520600750653

39 **'so as to humiliate him through defeat':** TNA ADM 223/84

39 **'a period of horoscopes, crystal-gazing and guesswork':** Winter, P. R. J., 'Libra rising: Hitler, astrology and British intelligence, 1940–43', *Intelligence and National Security,* 2006 21:3, 394-415, DOI: 10.1080/02684520600750653

40 **near impossible for bombers to find their targets:** Clarke, Gregory, *Deflating British Radar Myths of World War II*, War College Series, 2015, p. 15

40 **'generally bad navigators',** and Dowding's views on Luftwaffe navigation: TNA Air 20:6020

40 **Like a teenager asked to tidy his room:** TNA DEFE 40/1

41 **On another occasion, later in the war when he ran a small department:** ibid

41 **'The search through the intelligence files at once':** CA RVJO B.83

42 **'Could you also tell me whether any intensive effort is being made':** TNA DEFE 40/1

42 **'not even a secretary':** ibid

42 **'dishevelled troops on their way to Victoria':** Jones, *Most Secret War*, p. 90

42 **'For myself I had no doubt':** From a lecture given by Jones to the CIA, 1947

43 **'a tiny, vivacious, brainy blonde',** and the accounts of hairy-handed nuns etc.: Crang, Jeremy and Addison, Paul, *Listening to Britain: Home Intelligence Reports on Britain's Finest Hour, May–September 1940*, p. 28

44 **'I conceived scientific intelligence, with its constant vigil':** TNA DEFE 40/1

45 **an intelligence service was pointless unless:** TNA Air 20/1719

45 **'It is regretted that no separate summary could be issued':** TNA DEFE 40/1

45 **'My failure to obtain help in the early days had one interesting result',** and subsequent philosophizing on the job: TNA Air 20/1719

47 **'suspicious incidents' around West Beckham:** CA RVJO B.8

47 **'It is not a nice thing to ransack someone else's house',** and more colour on the raid: Jones, *Most Secret War*, p. 116

4. *The Clues*

49 **When his captors arrived, the German airman**, and subsequent account of the airman's notebook: CA RVJO B.7

50 **Jones liked German airmen**: CA RVJO J.1

50 **German-speaking interrogation officers were posted at RAF bases across the country**: TNA Air 40/1177

51 **'It must be remembered that an interrogator can only use words'**: ibid

52 **'London was not lying in ruins as they'd been led to believe'**: ibid

52 **'some were uncertain as to the limits of such politeness'**: ibid

54 **'You may remember that when Joshua attacked Jericho'**: TNA Air 20/1719

54 **'Information which has had to jump the gap between the sexes'**: Bristol Archives, DM1310 J112

54 **'Usually the statements obtained from a drunken man'**: CA RVJO B.114

54 **'the flow of information which he continuously maintained'**: TNA DEFE 40/21

54 **'not enough work to justify the employment of two people'**: Jones, Reginald, 'RAF Scientific Intelligence', *Air Intelligence Symposium Bracknell Paper No. 7*, 1997, p. 18

54 **'could only be a radio apparatus'**: TNA WO 208/3506

56 **'I naturally did not let him notice'**, and other prisoner transcripts: ibid

57 **the evidence from a downed Heinkel III**: TNA DEFE 40/2

58 **'We should present him with Wilhelm Busch's book Hans Knickebein'**: TNA WO 208:3506

59 **'And of course, that was it'**: CA RVJO K.475

5. The Chase

60 "'I'm very sorry, I can't say anything'": WO 208/3506

64 Later, captured documents showed: TNA Air 41:46

64 'in the nature of freaks': TNA Air 20/6020

64 'Mr. T. L. Eckersley, who is a world authority': CA RVJO B.7

67 'The excellence of these observations': ibid

67 'Until you've tried it, you have no idea': CA ROCO 1/1

68 'I was ushered into the great man's office': ibid

69 to see where Goering would turn his gaze for his next target: CA RVJO B.158

70 'Go to Moss Bros and hire a flight lieutenant's uniform': TNA Air 41/46

71 to form a 'jamming screen': ibid

71 'we can immediately jam the ether': TNA Air 20/6020

6. Cornish Patsy

72 The skies above Cornwall could be a lonely place, and the account of Dolenga's ill-fated flight: TNA Air 41/46

76 After the war, Johannes Plendl: CA B.83

76 'The crews using Knickebein soon reported': TNA A. D. I. (K) 334/1945

78 'Even when the enemy begins to suspect trouble': CA RVJO B.158

78 midway through a raid two Knickebein beams swapped frequencies: TNA Air 41/46

78 'obliterate the weak beams': ibid

79 'Presumably, during the intervening period': ibid

79 'The searchlights are close beside the guiding beam': TNA WO 208/3507

80 'Intelligence has a natural habit': CA RVJO B.23

80 'This particular aircraft had': TNA Air 20 6037/8792

81 'unholy glee at the alarm and despondency': Clayton, Aileen, *The Enemy is Listening*, p. 59

81 'no one yet dared admit the failure of Knickebein': TNA Air 41/10

81 'Now he wants to know what Knickebein is': TNA WO 208/3507

82 'We've got something': Jones, *Most Secret War*, p. 135

83 'To measure positions to this accuracy': CA RVJO B.16

7. The Spectre of the Brocken

85 In one of the surviving copies: CA RVJO B.18

86 'silly man': University of East Anglia archives, SZ/2/GEN

86 'Despite any unpopularity': Jones, *Most Secret War*, p. 139

86 each night he was advising Fighter Command: CA RVJO B.342

87 'You must keep flying to the left': CA WO 208/3507

88 'to reassure himself that he was right in not accepting religious dogma': Jones, *Sir Charles Frank, OBE, FRS: An Eightieth Birthday Tribute*, p. 5

89 'I wrote to him suggesting that he might be able to give me a better description': CA RVJO B.5

89 'It required very little distortion': ibid

91 this was the 'coarse' beam: CA RVJO B.18

95 'On some previous occasions': CA RVJO B.16

95 'Are we not tending to lose our sense of proportion over these German beams?': TNA DEFE 40/2

96 'Long may the Boche beam upon us': ibid

8. On Chesil Beach

97 Ostermeier was having a bad month: CA RVJO B.18

99   'We must be British and stick it': Crang, Addison, *Listening to Britain: Home Intelligence Reports on Britain's Finest Hour, May–September 1940*, p. 380

99   families 'making for the hopfields of Kent': ibid, p. 410

100   'It was the most amazing, impressive, riveting sight.': Clapson, *The Blitz*

101   released 'for some extremely important and urgent work': University of Bristol archives, DM1310/A.12

101   'a friendly alien, but an alien all the same': CA RVJO B.118

103   The noise, they wrote, 'brought a shock of positive pleasure'.: Crang, Addison, *Listening to Britain: Home Intelligence Reports on Britain's Finest Hour, May–September 1940*

104   'the most distinguished Gruppe in the German Bombing Force'.: CA RVJO B.18

104   The estimate, continued Jones: ibid

107   Here came the most cunning deduction of all: ibid

108   the radio was being interfered with by 80 Wing's meacons: TNA Air 41/46

109   'as grimly humorous as the Porter's performance in Macbeth': Jones, *Most Secret War*, p. 146

109   'one of these unfortunate wrangles': CA RVJO K.475

111   'we die together as a family': '*Mastermind* specialist subject – the Coventry Blitz', *The Coventry Telegraph*, 14 November 2008

112   'five hundred people are dead in the morning': CA RVJO K.475

112   'an uneasy night': Jones, *Most Secret War*, p. 103

113   'We knew this one would be different because it started early': '*Mastermind* specialist subject – the Coventry Blitz', *The Coventry Telegraph*, 14 November 2008

113   Before midnight, the fight was lost: 'Coventry Cathedral', feature on Historic England website, https://historicengland.org.uk/whats-new/features/blitz-stories/coventry-cathedral/

113 'The crew just gazed down at the sea of flames': Bekker, Cajus, *The Luftwaffe War Diaries*, p. 180

114 'There were more open signs of hysteria': Harrisson, Tom, Mass Observation Unit report

114 'very sharp . . . very accurate': CA RVJO K.475

115 'I express to the C.O. and this Gruppe my sincere thanks': Jones, *Most Secret War*, p. 166

9. Operation Starfish

116 Colonel John Turner was going through his own postbag: The account of Turner and the Starfish site draws heavily on Dobinson, Colin, *Fields of Deception*

117 became a passable impression of a burning city: TNA Air 41/46

118 the complexity of the radiation pattern: CA RVJO B.158

119 'literally thrown to the winds': CA RVJO B.23

120 'a Yorkshire moor rather than the factories of Derby': Clayton, Aileen, *The Enemy is Listening*, p. 76

120 'We heard a noise that is unforgettable': Jones, *Most Secret War*, p. 156

121 'and the sirens went to let us know the air raid was starting', and the rest of Corrigan's story: Plythian, Graham, *Manchester at War 1939–45*, p. 41

122 'The blast had lifted the whole lead roof of the cathedral up': ibid

122 'a long line of refugees coming to Prestwich': ibid

123 'extremely accurate in line, but had an error of one mile in range': TNA Air 41/46

123 interference to the approach beam was 'very serious': ibid

124 'It was usually possible to follow any such changes in the frequency within three or four minutes': CA RVJO B.18

124 'I asked one radio operator how much trouble': Price, Alfred, 'A new look at "the Wizard War"'

125 'Despite the good results obtained with "X", anger against the system': A. D. I. (K) 334/1945

125 'By April, the bulk of the Luftwaffe was generally being employed on moonlit nights': CA RVJO B.23

126 'If this be the battle of Manchester then Hitler has lost it': Plythian, *Manchester at War*, p. 8

10. *The King of the Gods*

128 'the German Air Force intended to celebrate the occasion in traditional style': Jones, Reginald, 'A tune on the Moothie', *New Scientist*, 20 December 1973

129 'It is proposed to set up Knickebein': Johnson, Brian, *The Secret War*, 2004

133 fallen with 'abnormal accuracy': CA RVJO B.23

133 Sometimes, the answers came literally fluttering down: CA RVJO B.20

134 'These changes suggested tension': CA RVJO B.23

135 'beset by quarrels in high places': TNA Air 41/10

135 'The squabble went on through September and October': CA RVJO B.23

138 'The internecine strife which occasionally developed was most heartening': CA RVJO B.158

138 'Night fighters, until airborne radar was good enough': CA RVJO K.475

139 'Was he really as good as he makes out': Bristol University Archives, DM1310

PART 2: ATTACK

11. *On a Plate*

145 **Langsdorff had no choice but to head for the nearest port**, and the account of the battle: 'The Battle of the River Plate', at New

Zealand's Torpedo Bay Naval Museum, https://navymuseum.co.nz/explore/by-themes/world-war-two-by-themes/the-battle-of-the-river-plate/#_ftn40

146 'This brilliant sea fight': Imperial War Museum https://www.iwm.org.uk/memorials/item/memorial/75701

147 'One teaspoonful of bicarbonate of soda per head per day', and other descriptions of the exchange: Winterbotham, Frederick, *The Ultra Spy*, p. 166

147 'How are you getting on with your experiments': Kinsey, Gordon, *Bawdsey*, p. 45

148 'was manned and operated day and night', and the account of Bainbridge-Bell's trip: CA RVJO B.159

149 'For a captain with a sense of honour, it goes without saying': Miller, David, *Command Decisions: Langsdorff and the Battle of the River Plate*, p. 159

12. Heligoland Bites

151 'The Tommies are not such fools – they won't come today': Bekker, Cajus, *The Luftwaffe War Diaries*, p. 71

151 That morning he had left his wife Mary at home, and the detailed accounts of the raid from the British side: Hastings, Max, *Bomber Command*, pp. ix–xxix

152 It would also drop propaganda leaflets over cities: TNA CAB 66/1/1

152 'inevitably involve incidental loss of life to civilians': TNA CAB 66/1/19

153 'You're plotting seagulls!': Bekker, *The Luftwaffe War Diaries*, p. 74

153 'Some amazing stories of the opportunities forgone by Great Britain': Hastings, *Bomber Command*, p. 19

155 'maintained that in the raids on Wilhelmshaven': CA RVJO B.24

155 '"direct electric rays"': CA RVJO B.12

155 'the Atlas works of Bremen were experimenting': CA RVJO B.6

155 'operates on much shorter wavelengths': CA RVJO B.13

155 'found himself in trouble with the Gestapo': CA RVO B.5

156 'because the special British wireless stations': CA RVJO B.12

156 'It was obvious that Freya represented something new': CA RVJO B.24

156 'The example of Wotan left little doubt of the association': CA RVJO B.8

158 'RDF stations fell intact into the hands of the enemy in France': Price, *Instruments of Darkness*, p. 74

13. *Proof Incontrovertible*

159 'To discuss the existence of German radar': Jones, *Most Secret War*, p. 191

159 There was Freya, of course: CA RVJO B.24

160 'As *Delight* had never been nearer than 60 miles': CA RVJO B.24

160 'there remains some expert prejudice against believing the Germans had radar': CA RVJO B.83

161 A large part of his subsequent success comes down: Bekker, *The Luftwaffe War Diaries*, p. 209

162 'From the counterintelligence viewpoint': Betts, T. J., 'Operation Columba', *CIA Historical Review Program*, released 22 September 1993

163 'these represented the minimum number to give continuous coastal cover': CA RVJO B.24

165 'An interpreter is like a motorist driving through a town': Downing, Taylor, *Spies in the Sky*, p. 89

165 'Interpretation of wireless apparatus': Medmenham Collection, MHP 494

168 'It was a gross disappointment': ibid

168 'A more foolhardy request it would be hard to imagine': Jones, *Most Secret War*, p. 190

168 not unreasonably, he was arrested as a spy: ibid, p. 191

14. *The Flight of the Hornet*

170 A lifetime later: Ryan, Mark, *The Hornet's Sting*, p. 104

171 'one of the most spectacular pieces of intelligence': CA RVJO B.24

172 'the capability to plot the position of a ship': Ryan, *The Hornet's Sting*, p. 22

173 'magic mirror': CA RVJO B.37

173 a 300m-high wireless mast: CA RVJO B.113

173 'At the same time I pulled down my trousers and pants': Ryan, *The Hornet's Sting*, p. 41

175 'he would kill as many Germans as he could': ibid, p. 70

176 'security was sacrificed to simplicity': CA RVJO B.24

177 'In the purely air offensive': CA RVJO J.1

177 'There is one thing that will bring him back and bring him down': https://winstonchurchill.org/publications/finest-hour/finest-hour-137/churchill-proceedings-churchill-and-bombing-policy/

177 'Movement, it must be remembered': Medmenham Collection MHP 494

179 'We shall need to know everything we can about the German night defences': Clark, *The Rise of the Boffins*, p. 171

179 'There was an inevitable irony about such episodes': Jones, *Most Secret War*, p. 201

179 'I went mad when I realized what happened': Ryan, *The Hornet's Sting*, p. 103

179 broadcasting at a wavelength of 50cm: CA RVJO B.24

181 'so small that several photographs had to be examined': TNA DEFE 40/2

181 'It was all too easy to ask for photographs': Jones, *Most Secret War*, p. 223

182 'When you pilots follow down the French coast': Medmenham Collection MHP 494

182 'He thereupon taxied his aircraft over to the others': Clark, *The Rise of the Boffins*, p. 176

15. *The Bruneval Raid*

184 **There were many things Major John Frost:** The account of the Bruneval Raid draws on and joins together numerous sources, primarily the original accounts in the National Archives, reference TNA DEFE 2/100–104, AIR 2/7689, AIR 20/1719; the popular account of the raid, Millar, George, *The Bruneval Raid*. Damian Lewis's *SAS Shadow Raiders* helped me keep the timeline in the correct order, as well as pointing me to sources.

191 **mocked up by Basil Schonland:** Austin, Brian, *Schonland: Scientist and Soldier*

16. *The Kammhuber Line*

196 **the Germans could now 'see' the bombs in the planes:** Jones, *Most Secret War*, p. 116

197–8 **'he still believes that the instrument "sees"':** CA RVJO B.146

198 **'corresponding ability to deceive':** TNA DEFE 2/102

198 **In January he noted to his wife:** ibid

200 **'an entangled pattern of barbed wire exceeding in complexity anything yet seen':** ibid

201 **It passed checkpoints and borders:** CA RVJO B.113

201 **showing the position of the night defences:** CA RVJO J.1

201 **'daunted by the prospect of cycling':** Jones, *Most Secret War*, p. 267

202 **two-thirds being downed by fighters:** CA RVJO B.157

202 **'New boxes began to spread ivy-fashion':** CA RVJO J.1

203 **'In view of the apparent current interest in similar installations':** Jones, *Most Secret War*, p. 269

203 **The policy at the time was to leave countermeasures until as late as possible:** CA RVJO B.113

204 **the most detailed piece of intelligence they had:** ibid

204 **they had fleshed out the entire defences:** CA RVJO J.1

205 **'a natural concept for them to regard':** CA RVJO B.57

206 **getting to within 20 to 50 metres:** CA RVJO B.37

206 **'Some pilots would take advantage of a time':** CA RVJO B.157

207 **executed in the Belgian city of Halle:** Bristol University Archives, DM1310

17. Radar Goes Mobile

208 **'This is the turning point of the war':** Lovell, Bernard, *Echoes of War*, p. 94

209 **'no one could shift him':** Bernard Lovell's diaries, given to the author

210 **'the most valuable cargo ever brought to our shores':** Stanley, Will, '1940s: The Cavity Magnetron', The Science Museum blog

210 **'Now the lab coat was replaced by a bulky flying suit':** Lovell, *Echoes of War*, p. 17

213 **they would risk H2S missions over enemy territory:** ibid, p. 151

214 **'"fingers of bright light sticking out into the darkness of the Elbe"':** ibid, p. 152

216 **'they had in the receiver and prisoners enough data to unravel the Gee system':** CA RVJO B.42

217 **'You must admit that at any rate we now have the "J" beams to get us to our targets':** ibid

217 'a marvellous opportunity' to apply 'as much of the national resources at my disposal': Jones, *Most Secret War*, p. 219

217 'Almost every step taken to suppress Gee': CA RVJO B.42

218 Josef Schmid still referred – apparently wholly unaware of the dupe: Isby, David, *Fighting the Bombers*, p. 155

219 Whether pilot or puss sets the pace: Clark, *The Rise of the Boffins*, p. 189

18. 'Emil-Emil'

223 'Ergo, you must find a place for the eye': Bekker, *The Luftwaffe War Diaries*, p. 214

224 'It meant that the romantic age of flying was past': ibid

224 He would get forty-four kills before being brought down: ibid, p. 303

225 'Their average life,' one document explained, 'was not long enough for their efficiency to reach a high level': CA RVJO B.158

226 'The last major gap in our knowledge of German night defences': CA RVJO B.83

227 'Majestically and calmly they move through the turbulent sea': 'A stirring memory' – the anniversary of the Channel Dash, Royal Navy blog, February 2021

229 'discussed with him all the ways we would set to work to fox his baby': Clark, *The Rise of the Boffins*, p. 63

229 'One is tempted to believe that there was an unconscious': CA RVJO B.158

229 'persuading him that you are either (a) where you are not, or (b) not where you are': CA RVJO B.24

230 'The chief tactical advantage gained by the present use of Mandrel': CA RVJO B.47

230 'Three fighters fitted with Moonshine caused the entire enemy fighter force': CA RVJO B.158

231 Were our radio communications resilient enough?: ibid

19. Open the Window

232 **Buck Ryan's face was ashen:** Price, *Instruments of Darkness*, p. 118

233 **a BBC engineer who had, before the war, proposed dragging a wire mesh:** TNA Air 20:6020

234 **sitting in her kitchen she began cutting up strips of foil:** Dee, Philip Ivor, *Biographical Memoirs of Fellows of the Royal Society*, 30 November 1984, pp. 139–66

235 **hastily dispatch police across much of the county to pick up the top-secret tin foil:** Clark, *The Rise of the Boffins*, p. 200

235 **'When the false responses are as numerous as the stars in the sky':** CA RVJO B.39

236 **'our own radar equipment was probably more susceptible to a Window attack':** CA RVJO B.158

237 **'We dared not experiment with the little beasts':** TNA A. D. I. (K) 334/1945

237 **'Open the window':** Clark, *The Rise of the Boffins*, p. 202

237 **resolutely refused to waggle:** CA RVJO B.49

237 **'The bombers are multiplying!':** ibid

238 **'Nightfighting, as well as the aimed and directed fire of the flak, was completely paralysed':** Isby, *Fighting the Bombers*, p. 75

239 **'The howling firestorm forced us to use our last strength':** *The Hamburg Police President's Report on the Large Scale Air Attacks on Hamburg, Germany, in World War 2*, Appendices, p. 88

240 **'"torrent of paper"':** CA RVJO B.113

240 **'Many German nightfighters would spend fruitless hours':** CA RVJO J.1

241 **'By this means the enemy has delivered the long awaited blow':** ibid

241 **'Hamburg was wrecked that night':** ibid

20. *Feints and Counterfeints*

242 **On the way down, he dropped a small leather bag:** CA RVJO B.83

243 **'all scruples and restrictions had been blasted':** CA RVJO B.113

243 **it sounded like bagpipes:** ibid

244 **'The use of women's voices proved to be futile':** Isby, *Fighting the Bombers*, p. 142

244 **'We had a woman waiting too, and so confusion was immediately restored':** CA RVJO J.1

244 **'transmissions of excited debates in Parliament':** Isby, *Fighting the Bombers*, p. 140

245 **the RAF designed Serrate:** CA RVJO B.158

245 **'The very device which was intended to enable our bombers':** ibid

245 **Grocer and Airborne Grocer, Cigar, Jostle and Piperack, Bagful, Blonde and Boozer:** ibid; TNA Air 8/831

245 **the German scientist Professor Leo Brandt took it up to the roof of a tall building:** Clark, *The Rise of the Boffins*, p. 197

247 **IFF did confuse searchlights, they insisted:** Jones, *Most Secret War*, p. 210

247 **'Our whole stream was thus betrayed, and its intentions divined':** CA RVJO J.1

248 **'the pathfinders leading the squadrons were completely under surveillance':** Clark, *The Rise of the Boffins*, p. 198

248 **'they spent nights searching the ether for new frequencies':** CA RVJO B.343

248 **'The enemy and ourselves had mustered considerable homing effort':** CA RVJO B.158

249 **'Only with difficulty could we restrain ourselves from entering':** CA RVJO B.70

249 **adding an aural humming sound to the visual echo:** CA RVJO B.58

249 'All technical progress was of no use anymore – they had come too late': Isby, *Fighting the Bombers*, p. 133

249 'in the final outcome the enemy's defences were completely neutralized': CA RVJO B.158

21. D-Day

251 'anticipating the reopening of the First Front': CA RVJO B.61

251 'It was arranged to attack two stations outside the landing area for every one within it': CA RVJO J.1

252 They had been learning to fly circuits: CA RVJO B.157

253 'Once a broad arrow representing an enemy attack appeared': Price, 'A new look at "the Wizard War"'

254 'It was clearly necessary to prevent him from making a successful interception': CA RVJO B.158

254 '1,000 Allied transport aircraft that were on their way to drop paratroops': Furner, Jack, 'Confound and . . .', *Royal Air Force Historical Society, Journal 28*, 2003

256 'As a final pinprick, General Schmid would castigate': CA RVJO J.1

256 'Frustrated, tormented and thrashed, the German night fighters never recovered': CA RVJO B.61

Epilogue

258 'Flight Officer Masterman has been showing repeated signs of conviviality at the slightest excuse': CA RVJO B.125

259 'A friend of yours was here last week': Jones, *Most Secret War*, p. 516; University of Oxford, Nuffield College, CHER D.124/15

Bibliography

Austin, Brian, *Schonland: Scientist and Soldier*, 2002, CRC Press

Bekker, Cajus, *The Luftwaffe War Diaries*, 1966, Ballantine

Bensusan-Butt, David, 'The Butt Report', https://etherwave.files.wordpress.com/2014/01/butt-report-transcription-tna-pro-air-14-12182.pdf

Churchill, Winston, *The Second World War Volume 2: Their Finest Hour*, 1949, Penguin Classics

Clapson, Mark, *The Blitz Companion, Air Raids in Britain: Aerial Warfare, Civilians and the City Since 1911*, 2019, University of Westminster Press

Clark, Ronald, *The Rise of the Boffins*, 1962, Phoenix House

Clayton, Aileen, *The Enemy is Listening*, 1980, Hutchinson

Crang, Jeremy, Addison, Paul, *Listening to Britain: Home Intelligence Reports on Britain's Finest Hour, May to September 1940*, 2011, Bodley Head

Dee, Philip Ivor, *Biographical Memoirs of Fellows of the Royal Society*, 30 November, 1984, Royal Society

Dobinson, Colin, *Fields of Deception*, 2000, Methuen

Downing, Taylor, *Spies in the Sky*, 2011, Abacus

Fry, Helen, *The London Cage: The Secret History of Britain's World War II Interrogation Centre*, 2017, Yale University Press

Goodchild, James, 'R. V. Jones and the Birth of Scientific Intelligence', PhD thesis, March 2013

Hastings, Max, *Bomber Command*, 1979, Zenith Press

Hinsley, Harry, *British Intelligence in the Second World War*, 1993, Cambridge University Press

Isby, David, *Fighting the Bombers*, 2003, Frontline Books

Jones, Reginald, *Most Secret War*, 1978, Coronet Books, Hodder & Stoughton

Jones, Reginald, *Reflections on Intelligence*, 1989, Mandarin

Judkins, Phil, 'Making Vision into Power', PhD thesis, 23 May 2008

Kinsey, Gordon, *Bawdsey*, 1983, Terence Dalton

Lewis, Damien, *SAS Shadow Raiders*, 2019, Quercus

Lovell, Bernard, *Echoes of War*, 1991, CRC Press

Millar, George, *The Bruneval Raid*, 1975, Doubleday & Co.

Miller, David, *Command Decisions: Langsdorff and the Battle of the River Plate*, 2013, Naval Institute Press

Plythian, Graham, *Manchester at War 1939–45*, 2014, The History Press

Price, Alfred, *Instruments of Darkness*, 1967, HarperCollins

Price, Alfred, 'A new look at "the Wizard War"', *Royal Air Force Historical Society Journal*, 28, 2003

Rowe, Albert, *One Story of Radar*, 1948, Cambridge University Press

Ryan, Mark, *The Hornet's Sting*, 2009, Skyhorse Publishing

Various, *The Rise and Fall of the German Air Force 1933 to 1945*, 1948, reissued 2008, A&C Black Business Information and Development

Winterbotham, Frederick, *The Ultra Spy*, 1989, Pan Macmillan

Acknowledgements

Because of Ramsey Faragher, a modern radio researcher, this book exists. I thank him for the inspiration, and for his generous technical proofreading.

I also thank Brian Austin. With a book like this it is hard to find someone with the skills to read the science, the knowledge to understand the history and the willingness to point out mistakes. Brian combines all three. His services came to me through serendipity.

One day, late in this book's writing, in my day job as science editor at *The Times*, I wrote a news story about moth migration. Scientists had tracked the moths' route over the Alps, on an epic journey along the spine of Italy. They had followed them thanks, I wrote, to radio transponders. Even as I wrote the words, I made a mental note to check them. I never did, but the next morning over breakfast Brian certainly did. I like to imagine him spluttering when he saw them. Surely I meant transmitters, he asked, in an email winningly titled 'Terminological Inexactitude'.

I like to think my subconscious had set a trap. My prey was that rarest of creatures: a radio historian with an eye for mistakes. Brian, I realized, was the author of *Schonland: Scientist and Soldier*, a 2002 biography of the South African radio scientist Sir Basil Schonland. Having corrected me once on an article, I asked Brian if he would do so for an entire book. He kindly agreed.

Any mistakes, obviously, are my own.

This book was written in part during a pandemic. Across Britain, libraries were shut. Without the extremely kind assistance of the team at the Churchill College Archives centre, who scanned and emailed copies from Jones's archives, the book would have taken many months longer.

I would particularly like to thank Andrew Riley, Chris Knowles, Madelin Evans, Jess Collins and Kath Thomson, all of whom did far more than they needed to. I owe Churchill College itself thanks for a lot more than just diligent copying.

Other archivists were equally kind, notably Ruth Pooley and Tim Fryer at the Medmenham Collection. Simon Fowler's ability to keep the copies coming from the National Archives through much of the pandemic also speeded the book's passage to deadline.

The team at Transworld have been a joy to deal with. Eloisa Clegg, like an Oboe beam across the dark of occupied Europe, directed the book to its target. Stephanie Duncan skilfully opened the bomb bay doors. To extend the metaphor far longer than I should, the exacting editorial eyes of Barbara Thompson, Katrina Whone, Steve Dobell, Elizabeth Dobson and Richard Mason ensured everything was air-worthy, while Tom Hill and Sophie MacVeigh in marketing and publicity are there to ensure the radio messages get out to the outside world.

To definitively overextend the metaphor, Phil Evans, who dealt with production, has been the armaments minister for the book, while Phil Lord, who commissioned the beautiful illustrations, and Richard Shailer, who designed the beautiful cover, have ensured it reaches its target in style.

Once again, Sarah Williams, my agent, has been there whenever I needed her. Knowledgeable and calm, she efficiently cut through the chaff (or Window) of the publishing world to show me the way.

Practically, my parents provided invaluable childcare for writing time – although suggesting that is all they did to help, in all manner of ways, would be insultingly inadequate. Professionally, I owe almost everything to *The Times*. Personally, I have relied on the love and crit-ical eye of Catherine. She was my first reader. My first listeners were Felix and William – if a book can be enjoyed by a six- and eight-year-old, then there is hope for it. Christopher will get there eventually.

Picture Acknowledgements

Page 14
Portrait of R. V. Jones: Courtesy of Aberdeen University.

Pages 10–11
Two Dornier DO617 bombers of the German Luftwaffe flying over the Royal Victoria Docks and Silvertown, West Ham, in the London suburbs, 7 September 1940: Photo12/Universal Images Group via Getty Images.

Pages 140–1
The *Graf Spee*, scuttled in the River Plate estuary, December 1939: Shawshots/Alamy Stock Photo.

Page 164, bottom
German 'Wurzburg C' anti-aircraft radar with parabolic reflector designed by Telefunken. 1940: Photo12/Universal Images Group via Getty Images.

Page 183
RAF reconnaissance photograph by Squadron Leader Anthony Hill, No. 543 Squadron RAF, of the Würzburg radar system on the cliffs at Bruneval, 5 December 1942: Pictorial Press Ltd/Alamy Stock Photo.

Page 205
Radar plotting table: The National Archives, ref. AIR40/3023.

All other diagrams and maps: Global Blended Learning Ltd.

Index

About the Author

Tom Whipple is the science editor at *The Times*. He covers everything from archaeology to zoology. He writes news, features, reviews and commentary across the paper, as well as appearing regularly on Times Radio. He joined the paper in 2006, shortly after graduating with a degree in mathematics.

During the course of his job he has visited the tunnels below Cern and the top of Mont Blanc above it. He has seen the inside of the world's hottest sauna and the world's most irradiated nature reserve. He has interviewed Stephen Hawking and Jedward. He has been arrested in three different countries.

As well as *The Times*, he has written for the *Guardian* and *The Economist*. He was named science journalist of the year for his coverage of the Covid-19 pandemic. This is his fifth book.